Legacy and Emerging Pollutants and Their Effects through the Lens of Environmental Management

Legacy and Emerging Pollutants and Their Effects through the Lens of Environmental Management

Guest Editors

Petros Samaras
Dorothea S. Kasiteropoulou
Christina Emmanouil

MDPI

Basel • Beijing • Wuhan • Barcelona • Belgrade • Novi Sad • Cluj • Manchester

Guest Editors
Petros Samaras
Department of Food Science and Technology
International Hellenic University
Thessaloniki
Greece

Dorothea S. Kasiteropoulou
Department of Environmental Studies
University of Thessaly
Larisa
Greece

Christina Emmanouil
School of Spatial Planning and Development
Aristotle University of Thessaloniki
Thessaloniki
Greece

Editorial Office
MDPI AG
Grosspeteranlage 5
4052 Basel, Switzerland

This is a reprint of the Special Issue, published open access by the journal *Journal of Xenobiotics* (ISSN 2039-4713), freely accessible at: https://www.mdpi.com/journal/jox/special_issues/0436MTX5Z8.

For citation purposes, cite each article independently as indicated on the article page online and as indicated below:

Lastname, A.A.; Lastname, B.B. Article Title. *Journal Name* **Year**, *Volume Number*, Page Range.

ISBN 978-3-7258-4813-3 (Hbk)
ISBN 978-3-7258-4814-0 (PDF)
https://doi.org/10.3390/books978-3-7258-4814-0

Contents

About the Editors

Petros Samaras

Petros Samaras is a doctor of Chemical Engineering and professor of Water Technology and Environmental Protection at the Department of Food Science and Technology, International Hellenic University Thessaloniki, Greece. His research activities are related to activated sludge treatment of municipal and industrial wastewaters; advanced treatment methods using physical and chemical processes for the removal of biorefractories from high-strength industrial wastewaters; energy auditing in municipal wastewater treatment plants and energy recovery from effluents and wastes; the optimization of the operation of membrane bioreactors for minimum membrane fouling; N-recovery from high-nitrogen-content effluents; the utilization and valorization of microalgae for the treatment of wastewaters and production of high-added-value substances; the production and use of char and activated carbon with tailored properties for the adsorption of target compounds; and the monitoring of the environmental impacts of waste and wastewater processes. He has participated in/coordinated around 65 research projects funded by the EU and national institutions. Currently, he is the coordinator of the HORIZON RIA project titled "A novel compact and advanced hybrid microalgae-membrane photobioreactor optimized for retrofitting decentralized water and wastewater management systems-NAMOR" (2025-2029) and team leader of the HORIZON IA project "Accelerating the sustainable production of advanced biofuels and RFNBOs - from feedstock to end-use - FUELPHORIA" (2023-2027). His bibliometric data (scopus, 2025) include 148 documents, with an h-index of 38 and 4526 citations.

Dorothea S. Kasiteropoulou

Dorothea Kasiteropoulou is a doctor of Civil Engineering and an associate professor of Hydraulic Works Simulation at the Department of Environmental Sciences, University of Thessaly, Greece. Since joining the University of Thessaly, Dr. Kasiteropoulou has conducted research on hybrid atomistic continuum methods, micro- and nano flow and multiscale modeling, computational modeling through nanoporous materials, computational fluid dynamics problems, transport phenomena in micro- and nano-fluidic systems, high-performance computing, and GPU programming. She has supervised projects on computational methods for high Reynolds number external and internal flows, numerical grid generation (finite elements), the vectorization of computer codes, the modeling of complex environmental processes, and the development of parallel processing algorithms. Dr. Kasiteropoulou has extensive research experience in the areas of sustainable water resources management and the management of steel industry scrap soil by means of its utilization into red ceramics. She has participated in 23 research projects funded by the EU and other national institutions and co-authored 73 papers in peer-reviewed journals and at international conferences, in addition to one book chapter.

Christina Emmanouil

Christina Emmanouil has worked as a lecturer ("Environmental management and risk assessment for pollutants in urban and rural areas and areas for productive and business activities") at the School of Spatial Planning and Development, Faculty of Engineering, Aristotle University of Thessaloniki, Greece, since 2018. Her current research focuses on solid waste management, urban wastewater management, and the contribution of different land uses to air, water, and soil pollution. She has previously worked as an evaluator for Draft Registration Reports (Ecotoxicology), according to Directive 91/414/EEC and Regulation 1107/2009/EC, and Competent Authority Reports (Environment), according to Directive 98/8/EC, for the Greek Ministry of Agriculture, as well as a visiting lecturer at the University of Thessaly, University of West Macedonia, and International Hellenic University, at both undergraduate and postgraduate levels. She has participated as an investigator or principal investigator in more than 30 projects on environmental management and environmental protection at Aristotle University of Thessaloniki, University of Thessaly, and the Agricultural University of Athens; she has co-authored 52 papers in peer-reviewed journals, three book chapters, and two textbooks (in Greek); and has acted as a Guest Editor for 12 Editorials in international journals.

Journal of
Xenobiotics

Editorial

Legacy and Emerging Pollutants and Their Effects Through the Lens of Environmental Management

Christina Emmanouil [1,*], Petros Samaras [2] and Dorothea Kassiteropoulou [3]

[1] School of Spatial Planning and Development, Aristotle University of Thessaloniki, 541 24 Thessaloniki, Greece
[2] Department of Food Science and Technology, International Hellenic University, 570 01 Thessaloniki, Greece; samaras@ihu.gr
[3] Department of Environmental Sciences, University of Thessaly, 382 21 Volos, Greece; dkasiter@uth.gr
* Correspondence: chemmanouil@plandevel.auth.gr

In the quest for worldwide economic development, the principles of sustainable development that involve social justice and efficient environmental protection are sometimes forgotten. One aspect of this problem includes expansive industrialization and urbanization that introduce both legacy [1,2] and emerging pollutants [3,4] into our environment; these pollutants may threaten ecosystems and human health [5]. Addressing this crisis requires a deep understanding of the nature of these pollutants and the deployment of innovative removal strategies [6]. The available tools include sustainable practices, such as the circular economy, that maximize resource efficiency [7] and minimize waste generation [8]. Nevertheless, these novel circular business models of reuse and recycling can also aggravate the (bio)accumulation of well-known or emerging pollutants within a semi-closed ecosystem [9]; hence, this is a double-edged sword. These risks have become a topic of major concern, as chemical or biological agents are widely and increasingly used in many activities, resulting in their continued release and recirculation into the environment through several different routes. The "sinks" of these routes may include air (particulate matter (PM), volatile organic compounds (VOCs), and industrial emissions) [10–12], water and sediment (metals, microplastics, persistent organic pollutants (POPs), pharmaceuticals, personal care products) [13–16], and soil (metals, POPs, polyfluoroalkyl substances (PFAs)) [17–19], among others. Pollutants may exert distinct and profound effects on ecosystems. In terrestrial systems, they may disrupt soil microbiota and reduce soil fertility [20] and plant growth or they may bioaccumulate within higher animals affecting reproductive health and longevity. In aquatic ecosystems, they may interfere with the reproduction and nutrition of many species. A recent study revealed that 80% of marine protected areas are exposed to potentially harmful levels of pollutants [21], impacting food webs and biodiversity [22]. In this context, the present Special Issue explores the landscape of classic (legacy) and emerging pollutants, focusing on novel monitoring methods, as well as on less well-studied pollution impacts on organisms and ecosystems. Understanding pollutant behavior across environmental compartments, unraveling their detailed effects on ecosystems, and mitigating their impacts on life will be critical in safeguarding our planet's future.

In this regard, Campani et al. (Contribution 1) studied fungicide mixtures with multiple active ingredients that may affect bees differently than single active compounds in vineyard agroecosystems. This toxicological research used a multi-biomarker approach and revealed disruptions impairing vital bee functions. The limited available data on the sublethal effects of fungicides on bees are highlighted, requiring more comprehensive evaluation of these chemicals on pollinator health.

https://doi.org/10.3390/jox15030089

In similar terrestrial ecosystems, the exposure of earthworms (*Eisenia fetida*) to short-chain perfluoropropylene oxide acids resulted in significant biochemical and behavioral alterations, even at sub-micromolar concentrations, as shown in the study of Rotondo et al. (Contribution 2). The substance HFPO-DA seems to cause the most pronounced effects, representing a significant environmental contaminant. The ecological risks posed by both legacy and emerging PFAS are crucial, and further research into the long-term impacts on soil ecosystems is required.

In terms of freshwater ecosystems, the buildup of fecal contaminants, due to direct discharge of effluents from poultry farms, was verified in Silway River, Philippines, using advanced microbial quantification protocols (Contribution 3). Simple water parameters (pH, T and TSS level) can create favorable conditions for chicken fecal coliforms to thrive. These findings highlight the ecological risks posed by poultry waste and the need for improved waste management practices.

In terms of marine ecosystems, the exposure of the marine copepod *Tigriopus fulvus* to nine rare earth metals (REEs) demonstrated concentration-dependent increases in mortality and immobilization, with LC50 values ranging from 0.56 to 1.99 mg/L (Contribution 4). The exposure to REEs resulted in a significant slowing of naupliar development. Among the nine REEs, lanthanum proved to be the most toxic, with a high impact on the survival and development of *T. fulvus*, affecting copepod populations.

Burdened marine ecosystems, such as those in the Mediterranean, are highly worthy of investigation, as shown in the study of Kasiotis et al. (Contribution 5), who investigated both organic and inorganic contaminants in fish from the Thermaikos Gulf in Greece. Low concentrations of pharmaceuticals and pesticides were found in *D. labrax* and *S. solea*; however, trace metals were detected more frequently. The risk assessment indicated minimal health risks for adults. Nevertheless, the levels of mercury and arsenic could pose a threat to children. These findings emphasize the need for ongoing monitoring of seafood safety.

Furthermore, juvenile trout was used as a model organism to study the effects of graphene-based nanoparticles (Contribution 6). The exposure resulted in changes to their cardiorespiratory function, swimming behavior, and nesting activity. Graphene oxide enhanced swimming performance in juveniles by as much as 72% while reducing nesting by up to 23%. These alterations indicate potential developmental consequences on both behavior and physiology. Additional studies are required to explore the long-term ecological implications.

Subsequently, another model organism, *Daphnia magna*, was chosen to assess the effects of sublethal exposure to pyriproxyfen over a period of 21 days, in the study of Salesa et al. (Contribution 7). Notable biochemical changes included an increase in LDH activity and a reduction in cholesterol levels at a concentration of 14.02 μg/L, among others. These molecular and physiological changes imply that such biomarkers could act as early warning signals for pyriproxyfen.

Finally, the review paper of Ziegler (Contribution 8) summarizes the progress in toxicity testing utilizing duckweeds (Lemnaceae). Recent research has led to the development of fast protocols, the use of additional species, improved sensitivity, and higher throughput. The incorporation of omics technologies—transcriptomics, proteomics, and metabolomics—paired with FTIR spectroscopy and genotoxicity tests, has broadened the range of biomarker-based evaluations. These advancements allow for a more comprehensive understanding of toxic mechanisms and aid in the detection of unknown contaminants.

On the other hand, assessing environmental behavior rather than toxicity effects, Zhou et al. (Contribution 9) examined microparticle concentrations in the sea–surface microlayer and bulk water of Osaka Bay. Microparticle concentrations were elevated in the surface layer, primarily consisting of smaller particles. The predominant polymer identified was polymethyl methacrylate (PMMA). The findings indicate that antifouling paint from vessels is a significant source of pollution. These results underscore the influence of maritime activities on marine contamination.

Funding: This research received no external funding.

Institutional Review Board Statement: Not applicable.

Informed Consent Statement: Not applicable.

Conflicts of Interest: The authors declare no conflicts of interest.

List of Contributions

1. Campani, T.; Manieri, G.; Caliani, I.; Di Noi, A.; Casini, S. Apis mellifera as a Model Species to Evaluate Toxicological Effects of Fungicides Used in Vineyard Agroecosystems. *J. Xenobiot.* **2025**, *15*, 18. https://doi.org/10.3390/jox15010018.
2. Rotondo, D.; Gualandris, D.; Lorusso, C.; Braeuning, A.; Calisi, A.; Dondero, F. Impact of Short-Chain Perfluoropropylene Oxide Acids on Biochemical and Behavioural Parameters in *Eisenia fetida* (Savigny, 1826). *J. Xenobiot.* **2025**, *15*, 2. https://doi.org/10.3390/jox15010002.
3. Opog, L.M.; Casila, J.C.; Lampayan, R.; Sobremisana, M.; Bulasag, A.; Yokoyama, K.; Haddout, S. Assessment of Chicken Fecal Contamination Using Microbial Source Tracking (MST) and Environmental DNA (eDNA) Profiling in Silway River, Philippines. *J. Xenobiot.* **2024**, *14*, 1941–1961. https://doi.org/10.3390/jox14040104.
4. Biandolino, F.; Prato, E.; Grattagliano, A.; Libralato, G.; Trifuoggi, M.; Parlapiano, I. Potential Toxicity of Nine Rare Earth Elements (REEs) on Marine Copepod *Tigriopus fulvus*. *J. Xenobiot.* **2024**, *14*, 1919–1929. https://doi.org/10.3390/jox14040102.
5. Kasiotis, K.M.; Zafeiraki, E.; Manea-Karga, E.; Kouretas, D.; Tekos, F.; Skaperda, Z.; Doumpas, N.; Machera, K. Bioaccumulation of Organic and Inorganic Pollutants in Fish from Thermaikos Gulf: Preliminary Human Health Risk Assessment Assisted by a Computational Approach. *J. Xenobiot.* **2024**, *14*, 701–716. https://doi.org/10.3390/jox14020041.
6. Makaras, T.; Jakubowska-Lehrmann, M.; Jurgelėnė, Ž.; Šemčuk, S. Exploring the Effects of Graphene-Based Nanoparticles on Early Salmonids Cardiorespiratory Responses, Swimming and Nesting Behavior. *J. Xenobiot.* **2024**, *14*, 484–496. https://doi.org/10.3390/jox14020029.
7. Salesa, B.; Torres-Gavilá, J.; Ferrando-Rodrigo, M.D.; Sancho, E. Pyriproxyfen Contamination in *Daphnia magna*: Identifying Early Warning Biomarkers. *J. Xenobiot.* **2024**, *14*, 214–226. https://doi.org/10.3390/jox14010013.
8. Ziegler, P. Developments in Toxicity Testing with Duckweeds. *J. Xenobiot.* **2025**, *15*, 48. https://doi.org/10.3390/jox15020048.
9. Zhou, M.; Yanai, H.; Yap, C.K.; Emmanouil, C.; Okamura, H. Anthropogenic Microparticles in Sea-Surface Microlayer in Osaka Bay, Japan. *J. Xenobiot.* **2023**, *13*, 685–703. https://doi.org/10.3390/jox13040044.

References

1. Cantoni, J.; Kalantari, Z.; Destouni, G. Legacy Contributions to Diffuse Water Pollution: Data-Driven Multi-Catchment Quantification for Nutrients and Carbon. *Sci. Total Environ.* **2023**, *879*, 163092. [CrossRef]
2. Zhang, X.; Li, L.; Xie, Z.; Ma, J.; Li, Y.-F.; Cai, M.; Ren, N.-Q.; Kallenborn, R.; Zhang, Z.-F.; Zhang, X.; et al. Exploring Global Oceanic Persistence and Ecological Effects of Legacy Persistent Organic Pollutants Across Five Decades. *Sci. Adv.* **2024**, *10*, eado5534. [CrossRef]
3. Yu, Y.; Wang, Z.; Yao, B.; Zhou, Y. Occurrence, Bioaccumulation, Fate, and Risk Assessment of Emerging Pollutants in Aquatic Environments: A Review. *Sci. Total Environ.* **2024**, *923*, 171388. [CrossRef]
4. Donley, N.; Cox, C.; Bennett, K.; Temkin, A.M.; Andrews, D.Q.; Naidenko, O.V. Forever Pesticides: A Growing Source of PFAS Contamination in the Environment. *Environ. Health Perspect.* **2024**, *132*, 75003. [CrossRef] [PubMed]
5. Chen, F.; Jiang, F.; Ma, J.; Alghamdi, M.A.; Zhu, Y.; Yong, J.W.H. Intersecting Planetary Health: Exploring the Impacts of Environmental Stressors on Wildlife and Human Health. *Ecotoxicol. Environ. Saf.* **2024**, *283*, 116848. [CrossRef]

6. Akhtar, M.S.; Ali, S.; Zaman, W. Innovative Adsorbents for Pollutant Removal: Exploring the Latest Research and Applications. *Molecules* **2024**, *29*, 4317. [CrossRef] [PubMed]
7. Baldassarre, B. Circular Economy for Resource Security in The European Union (EU): Case Study, Research Framework, and Future Directions. *Ecol. Econ.* **2025**, *227*, 108345. [CrossRef]
8. Kotyal, K. Sustainable Waste Management in the Circular Economy: Challenges and Opportunities. *Environ. Rep.* **2023**, *5*, 1–5. [CrossRef]
9. Gonçalves Castro, C.; Hofmann Trevisan, A.; Pigosso, D.C.A.; Mascarenhas, J. The Rebound Effect of Circular Economy: Definitions, Mechanisms and a Research Agenda. *J. Clean. Prod.* **2022**, *345*, 131136. [CrossRef]
10. Sharma, R.; Kurmi, O.P.; Hariprasad, P.; Tyagi, S.K. Health Implications Due to Exposure to Fine and Ultra-Fine Particulate Matters: A Short Review. *Int. J. Amb. Energ.* **2024**, *45*, 2314256. [CrossRef]
11. Oh, S.H.; Choe, S.; Song, M.; Yu, G.H.; Schauer, J.J.; Shin, S.A.; Bae, M.S. Effects of Long-Range Transport on Carboxylic Acids, Chlorinated VOCs, and Oxidative Potential in Air Pollution Events. *Environ. Pollut.* **2024**, *347*, 123666. [CrossRef]
12. Emmanouil, C.; Drositi, E.; Vasilatou, V.; Diapouli, E.; Krikonis, K.; Eleftheriadis, K.; Kungolos, A. Study on Particulate Matter Air Pollution, Source Origin, and Human Health Risk Based of PM10 Metal Content in Volos City, Greece. *Toxicol. Environ. Chem.* **2016**, *99*, 691–709. [CrossRef]
13. Elumalai, P.; Parthipan, P.; Gao, X.; Cui, J.; Santhosh Kumar, A.; Dhandapani, P.; Rajasekar, A.; Sarma, H.; Vadivu Ganapathy, N.R.; Theerthagiri, J.; et al. Impact of Petroleum Hydrocarbon and Heavy Metal Pollution on Coral Reefs and Mangroves: A Review. *Environ. Chem. Lett.* **2024**, *22*, 1413–1435. [CrossRef]
14. Koelmans, A.A.; Mohamed Nor, N.H.; Hermsen, E.; Kooi, M.; Mintenig, S.M.; De France, J. Microplastics in Freshwaters and Drinking Water: Critical Review and Assessment of Data Quality. *Water Res.* **2019**, *155*, 410–422. [CrossRef] [PubMed]
15. Emmanouil, C.; Bekyrou, M.; Psomopoulos, C.; Kungolos, A. An Insight into Ingredients of Toxicological Interest in Personal Care Products and a Small–Scale Sampling Survey of the Greek Market: Delineating a Potential Contamination Source for Water Resources. *Water* **2019**, *11*, 2501. [CrossRef]
16. Vasseghian, Y.; Hosseinzadeh, S.; Khataee, A.; Dragoi, E.N. The Concentration of Persistent Organic Pollutants in Water Resources: A Global Systematic Review, Meta-Analysis and Probabilistic Risk Assessment. *Sci. Total Environ.* **2021**, *796*, 149000. [CrossRef]
17. Li, Y.F.; Hao, S.; Ma, W.L.; Yang, P.F.; Li, W.L.; Zhang, Z.F.; Liu, L.Y.; Macdonald, R.W. Persistent Organic Pollutants in Global Surface Soils: Distributions and Fractionations. *Environ. Sci. Ecotechnol.* **2023**, *18*, 100311. [CrossRef]
18. Golia, E.E.; Emmanouil, C.; Charizani, A.; Koropouli, A.; Kungolos, A. Assessment of Cu and Zn Contamination and Associated Human Health Risks in Urban Soils from Public Green Spaces in the City of Thessaloniki, Northern Greece. *EuroMediterr. J. Environ. Integr.* **2023**, *8*, 517–525. [CrossRef]
19. Wang, Y.; Munir, U.; Huang, Q. Occurrence of Per- and Polyfluoroalkyl Substances (PFAS) in Soil: Sources, Fate, and Remediation. *Soil Environ. Health* **2023**, *1*, 100004. [CrossRef]
20. Wołejko, E.; Jabłońska-Trypuć, A.; Wydro, U.; Butarewicz, A.; Łozowicka, B. Soil Biological Activity as An Indicator of Soil Pollution with Pesticides—A Review. *Appl. Soil Ecol.* **2020**, *147*, 103356. [CrossRef]
21. Abessa, D.M.S.; Albuquerque, H.C.; Morais, L.G.; Araújo, G.S.; Fonseca, T.G.; Cruz, A.C.F.; Campos, B.G.; Camargo, J.B.D.A.; Gusso-Choueri, P.K.; Perina, F.C.; et al. Pollution status of marine protected areas worldwide and the consequent toxic effects are unknown. *Environ. Pollut.* **2018**, *243 Pt B*, 1450–1459. [CrossRef]
22. Dumont, E.; Williams, R.; Keller, V.; Voß, A.; Tattari, S. Modelling Indicators of Water Security, Water Pollution and Aquatic Biodiversity in Europe. *Hydrol. Sci. J.* **2012**, *57*, 1378–1403. [CrossRef]

Journal of
Xenobiotics

MDPI

Article

Anthropogenic Microparticles in Sea-Surface Microlayer in Osaka Bay, Japan

Mi Zhou [1], Hirofumi Yanai [1], Chee Kong Yap [1,2], Christina Emmanouil [3] and Hideo Okamura [4,*]

[1] Graduate School of Maritime Sciences, Kobe University, Fukaeminami-machi, Higashinada-ku, Kobe 658-0022, Japan; zhoumi77@163.com (M.Z.); yahirofumi1996@yahoo.co.jp (H.Y.); yapchee@upm.edu.my (C.K.Y.)

[2] Department of Biology, Faculty of Science, Universiti Putra Malaysia, UPM Serdang, Serdang 43400, Selangor, Malaysia

[3] School of Spatial Planning and Development, Aristotle University of Thessaloniki, 54124 Thessaloniki, Greece; chemmanouil@plandevel.auth.gr

[4] Research Center for Inland Seas, Kobe University, Fukaeminami-machi, Higashinada-ku, Kobe 658-0022, Japan

* Correspondence: okamurah@maritime.kobe-u.ac.jp; Tel.: +81-078-431-6272

Abstract: The abundance, distribution, and composition of microparticles (MPs) in the sea-surface microlayer (S-SML, less than 100 µm of sea surface in this experiment) and in bulk water (1 m under the sea surface) were investigated to evaluate the pollution level of MPs in Osaka Bay in Japan. Both seawater fractions were collected at eight sites including ship navigation routes, the coastal area, and the center of Osaka Bay for 2021–2023. MPs were filtered for four size ranges (10–53, 53–125, 125–500, and >500 µm) and then digested with H_2O_2. MPs' abundance was microscopically assessed; and polymer types of MPs were identified by a Fourier transform infrared spectrometer (FTIR). For the 22 collections performed along eight sites, the average MPs' abundance was 903 ± 921 items/kg for S-SML, while for the 25 collections performed along the same sites, the average MPs' abundance was 55.9 ± 40.4 items/kg for bulk water, respectively. MPs in both S-SML and bulk water exhibited their highest abundance along the navigation routes. The smallest MPs (10–53 µm) accounted for 81.2% and for 62.2% of all MPs in S-SML and in bulk water among all sites, respectively. Polymethyl methacrylate (PMMA) was the major type of MPs identified while minor ones were polyethylene, polyesters, polystyrene, polypropylene, polyvinyl chloride, polyamide, etc. PMMA comprised 95.1% of total MPs in S-SML and 45.6% of total MPs in bulk water. In addition, PMMA accounted for 96.6% in S-SML and 49.5% in bulk water for the smallest MP category (10–53 µm). It can be assumed that the MP sources were marine paints—primarily APPs (antifouling paint particles)—as well as land coatings. Sea pollution due to microparticles from ship vessels should be given proper attention.

Keywords: microparticles (MPs); seawater surface microlayer; PMMA; antifouling paint particles (APPs); Osaka Bay

1. Introduction

The presence of microplastics (plastic particles less than 5 mm) is a ubiquitous problem in terrestrial and aquatic ecosystems. Once introduced into an aquatic environment by wind transport, stormwater runoff, drainage, and sewage system, plastic debris can be transported offshore and enter oceanic gyres by surface currents and winds, which leads to 80% of terrigenous plastic waste accumulating in a marine environment [1,2]. For instance, at least 79 thousand tons of ocean plastics are floating inside an area of 1.6 million km^2 at the Great Pacific Garbage Patch (GPGP), where microplastics account for 8% of the total mass [3]. Besides the residual of primary microplastics with microscopic sizes such as resin pellets and microbeads used in cosmetics and air-blasting media [4,5], secondary microplastics are continuously generated from the breakdown of larger plastic debris due to UV irradiation, mechanical forces and biodegradation by bacteria, fungi, and microalgae, and they also end up in marine ecosystems [6,7]. Compared to macroplastics, microplastics can exudate more toxic additives due to their large specific surface and their full exposure to seawater, and they can transmit hazardous chemicals to biota [8]. Additionally, microplastics are prone to adsorb persistent organic pollutants (POPs) such as PCBs and PBDEs from surrounding water medium [9]. Once these microplastics are ingested by all kinds of aquatic fauna including invertebrates, planktonic organisms, seabirds, crustaceans, fish, aquatic animal and marine mammals, they may bioaccumulate and biomagnify via the food web [10,11]. Microplastics have been discovered in a diversity of seafood items and even in human feces and blood clots [12,13]; as such, it can be assumed that humans are exposed to microplastics through their diet. This exposure may compromise fecundity and other somatic processes and potentially threaten human health [14].

It is assumed that 20% of marine-based plastic debris mainly stems from fishing gears, aquaculture, marine structures, and vessels [15]. Furthermore, antifouling paint particles (APPs) and other marine paint debris [15], can flake off from hulls during processes of shipbuilding, boat cleaning, and vessel navigation and then enter the local aquatic environment. This source has been incidentally or deliberately excluded from the characterization of microplastics due to the omission of paint in marine litter guidelines [16]. Nevertheless, some investigations have reported evidence of the occurrence of paint particles in seawater; paint microparticles in surface seawater accounted for 33% of all microplastics in the South China Sea [17] and 15.21% of all microplastics in the Jiangsu coastal area in China [18], respectively. In the sea-surface microlayer of the southern coast of Korea, floating alkyd resins (171 items/L) were predominant (81% of all microsized polymer particles) [11]. These alkyd resins may have originated from antifouling paint coatings of fishing boats; however, there was no additional evidence to prove this hypothesis. Furthermore, PMMA resins prevailed in the northwestern region of the German Bight with a concentration of 1032 ppt and accounted for 83% total polymer types. Given that the highest amount of PMMA was found within a major shipping lane, these PMMA resins were considered to originate from binders used in antifouling paint on ships [19]. Moreover, some microplastics discovered in European shelf seas in the North Atlantic Ocean were found to be Cu-based antifouling formulations, and other metals such as Pb and Fe therein were also related to antifouling paints from navigating ships [20]. In light of the higher chemical toxicity of paint particles than similarly sized microplastics of metallic additives [16], the investigation of APPs and marine paint particles is urgently needed, and it is the focus of the present study. Given these characteristics, herein, microplastics and paint microparticles including APPs are called microparticles (MPs).

The sea-surface microlayer (S-SML) is the interface between the atmosphere and sub-layer water. MPs in the S-SML can be transferred into the upper marine air through bubbles [21] or sink into the deeper water column or into sediments due to biofouling [22]. Meanwhile, due to the water surface tension and the sticky microgel produced by microbial activity [23], MPs are prone to accumulate in the S-SML. However, due to the relatively high requirements in time, cost, and equipment for collecting samples from the S-SML, compared with the collection of water column and sediments, investigations on MPs in the S-SML are few in Japan and worldwide. For all these reasons, this study focused on MPs in the S-SML that may yield extensive and severe MP pollution. It was noted in relevant investigations where manta trawl nets (330 μm mesh) or neuston trawl nets (300 μm mesh) were used for sampling that large-size MPs (>300 μm) were always detected because smaller MPs less than 330 μm could not be captured. For instance, MPs in surface seawater in the South China Sea were found to be around 0.045 ± 0.093 items/m^3 when a bongo net (333 um mesh) was used; however, this was changed to 2569 ± 1770 items/m^3 when a pump was used [24]. In the latter case, smaller-size fractions (<0.3 mm) contributed 92% of the total MPs load [24]. Even though tiny MPs (<50 μm) are significantly abundant [25] and pose a higher risk of being incorporated into organisms' tissues [26] than larger MPs, few studies identify and evaluate them [27]. Thus, these tiny MPs have been ignored in many studies due to high-tech instrument requirements and time-consuming determination [28], which can lead to a serious underestimation of MP loads in the environment and their corresponding risks [29]. As such, MPs of 10–53 μm size were especially collected and analyzed in this study.

Based on the above theories, we herein make the hypothesis that more MPs may accumulate in the S-SML and along navigation routes, and most of these MPs may be related to antifouling and marine paint. For evaluating the MP pollution status of Osaka Bay in Japan, the present study sought to (1) assess the occurrence and distribution of MP loads at different depths of the water column (S-SML and bulk water), (2) estimate the disparity of MP loads among different sampling zones such as navigation routes, the coastal area, and the bay center, and (3) investigate the abundance, spatial distribution, polymer types, size distribution, and sources of these MPs. This study can be of significance in providing data on MPs' fate and distribution at a regional or global scale, and it can also provide some theoretical guidance for controlling MP pollution.

2. Materials and Methods

2.1. Sampling Sites

Osaka Bay, located in western Japan, is connected to the enclosed sea area of Harimanada through the Akashi Strait in the northwest. Eastward water transport always prevails in the Seto Inland Sea due to westerly winds, and significant tidal currents with a complex dynamic mechanism are formed in the Akashi Strait, which leads to the contamination of Osaka Bay by the transport of pollutants through this strait [30]. Meanwhile, cyclonic circulation at Osaka Bay both in summer and in winter can also provide impetus to water conveyance from the Kii Strait in the southeast and finally, to the Pacific [31]. Coupled with domestic and industrial activities from international trade ports, large industrial centers, and coastal industrial cities in Hanshin Industrial Zone around Osaka Bay, this situation has led to the formation of terrestrial and marine plastic debris arising from the whole of Western Japan, which is then exported towards the Pacific.

It should be noted that the water circulation is quite significant in Osaka Bay with a tidal range about 1 m every day; as such, the MPs' abundance in the estuary is directly related to the tidal fluctuation [32]; for this reason, the estuary was excluded from the sampling sites as much as possible in this research, just apart from site 2 that locate at estuary of Yodogawa. Considering that the purpose of this study was to investigate the discrepancy in MPs' abundance among different marine zones, in total, the sampling sites included 3 sampling zones (Figure 1 and Table S1); S1, S2, and S3 are located in the coastal area, S4, S5, S6, and S7 are located along shipping navigation routes, and S8 is the center of Osaka Bay. This design allowed for the testing of our hypothesis that MPs' abundance along the navigation routes is significantly higher than that in the coastal area and the bay center.

2.2. Field Surveys

All water samples were collected from September to June from 2021 to 2023 (Tables S1 and S2). In order to more accurately reflect the average MPs' abundance at each sampling site, sampling was usually repeated 2–4 times. The repeated samplings were conducted on different dates, and it is impossible to ensure that the currents, tides, and meteorological conditions on these different dates were under the same level. In this sense, repeated samplings at the same site can yield results under different sea and meteorological conditions, which can reflect the average MPs' abundance with the local characteristics of Osaka Bay.

The S-SML was sampled by an electric rotating drum sampler (RDS) with an acrylic cylindrical drum (diameter 0.25 m, width 0.45 m, surface area 0.353 m^2). A total surface area of 212 m^2 of seawater was collected at 20 rotation per min for 30 min. Meanwhile, bulk water below 1 m of sea surface was sampled through a pump (SK-62510, Koshin, Kyoto, Japan) as a contrast substrate. Approximately 10 L of seawater samples (S-SML and bulk water) was collected for each site.

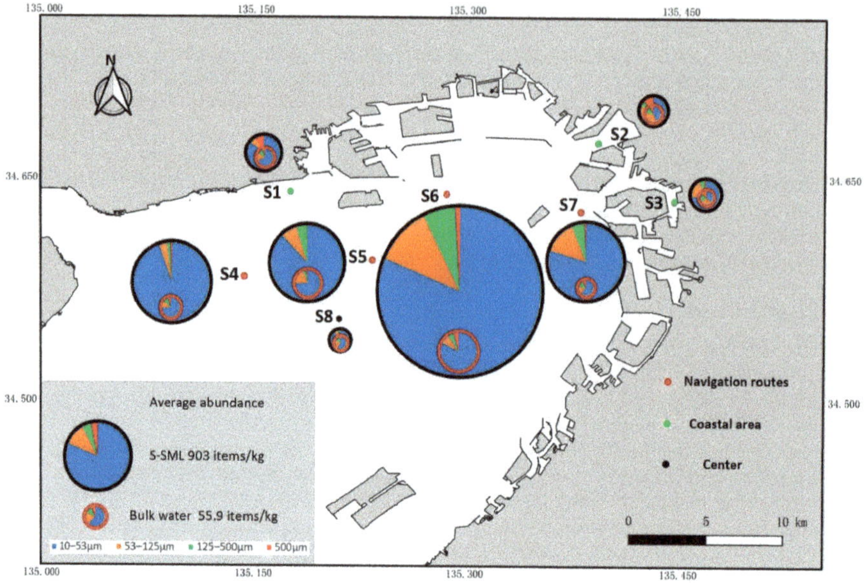

Figure 1. Sampling sites and spatial distribution of average MPs' abundance.

Both S-SML and bulk water samples were filtered through a stainless-steel sieve cascade (φ10 cm) (JIS Z 8801, Tokyo Screen, Tokyo, Japan). For the first filtration, the sieve was equipped with a 500 μm mesh-size filter, the subsequent filtration was performed with a 125 μm mesh-size filter, and the final filtration with a 53 μm mesh-size filter; then, the water was returned to a 10 L tank. The retained MPs on these sieves with separate size ranges of >500, 125–500, and 53–125 μm, respectively, were rinsed into bottles with the aid of ultrapure water. As a result of the sieving process, the remaining MPs in the tank water samples were less than 53 μm. Bottles and tanks were sealed and stored on the experimental ship, then carried back to the laboratory.

2.3. Laboratory Pretreatment

MPs in bottles (with size ranges of 53–125, 125–500, and >500 μm) were filtered through a PTFE filter (POPMF-1000, 10 μm, Wintec, Kobe, Japan) with a diameter of 47 mm, and MPs in tanks (with size of <53 μm) were filtered through a PTFE filter (POPMF-1000, 10 μm, Wintec, Kobe, Japan) with a diameter of 142 mm. The filters with retained MPs were kept in Petri dishes and dried at 60 °C for 12 h; then, they were soaked in 30% H_2O_2 solution for 1 week, to digest organic matter. For removing skins and shells of algae and microbe coverings on samples, MPs with size ranges of 10–53, 53–125, 125–500, and >500 μm on the filters were washed down and filtered through a nylon filter with a mesh size of 10 μm (#500/520, Tantore, Aichi, Japan), 48 μm (No. 305T, AS ONE, Osaka, Japan), 108 μm (No. 150T, AS ONE), or 500 μm (NB-34, AS ONE), respectively. Especially for MPs of size 10–53 μm, this vacuum filtering process lasted for more than 40 min to fully air-dry and fix algae shells on the nylon filter, so that MPs could be completely separated from them. The MPs retained on these nylon filters were rinsed apart from the fettered algae shells and filtered through a PTFE filter (H100A013A, ADVANREC, Osaka, Japan, 1 μm, 13 mmφ). Since MPs with a size of 10–53 μm could clog filter pores when in abundance, they were washed with 100 mL of ultra-pure water and filtered five times with five separate filters, rather than just filtered once. Finally, all filters were dried at 60 °C for 12 h and used for microscopic observations. All filtering processes operated in a vacuum filtration system. MPs were rinsed by ultrapure water, and filters were dried in a dryer (D-450 FA, AS ONE, Osaka, Japan). The relevant process of water sampling and laboratory treatment is shown in Figure S1.

2.4. Detection and Identification of MPs

MPs on filters were observed and counted by an infrared microscope (AIM-9000, SHIMAZU, Kyoto, Japan) and their polymer types were analyzed and identified by a Fourier transform infrared spectrometer (FTIR: IRTracer-100, SHIMADZU, Kyoto, Japan). For MPs with a size >500, 125–500, and 53–125 μm, a 1/4 square of the filter was chosen, the plasticlike particles retained there were detected, and MPs' abundances and corresponding polymer types were measured and recorded. For MPs with a size of 10–53 μm, due to their high concentrations and their laborious counting, the particles were counted in three squares of the filter paper only, each of which was randomly assigned from a center, middle, and outer zone of every PTFE filter [33]. One square was 1.6 × 1.8 mm^2, and three squares were approximately 1/9 of the total filtered area (Figure S2). MPs with a size of 10–53 μm on all five filters were counted repetitively in that way. The polymer type of each MP could be identified by measuring its transmittance and matching its IR spectrum to the one in the FTIR polymer spectrum library with a matching degree higher than 70%. All the polymer types found in this study were acrylic resins, mainly represented by polymethyl methacrylate (PMMA), polypropylene (PP), polyethylene (PE), polyester (PET), polystyrene (PS), nylon (PA), polyvinyl chloride (PVC), and others.

To exclude contamination from environmental MPs such as nylon filters, PS bottles, and PE tanks, ultrapure water was used as the blank water sample to repeat the above processes of treatment and filtration. No MPs were detected in the blank samples; as such, the contamination from the equipment was negligible.

2.5. Data Analysis

QGIS 3.38.0 (QGIS.ORG, https://qgis.org/ja/site/, accessed on 3 November 2023) was used for mapping sampling sites in the Osaka Bay shown in Figure 1. The differences in MPs' abundances in the S-SML and the corresponding bulk water below were assessed through the Mann–Whitney U test. In addition, the differences in MPs' abundance (containing all polymers or PMMA) in the S-SML or in bulk water among different sampling zones were analyzed by a one-way ANOVA followed by Tamhane's T2 post hoc test. A simple linear regression was carried out to test the linear relationship between MPs' abundance and size. A Pearson correlation analysis was conducted to measure the correlation between MPs' abundance and other factors including polymer types, size ranges of MPs and depth, pH, salinity, temperature, and TOC and DOC of water sampling. MP sources were summarized by the method of principle component analysis (PCA). The correspondence analysis was conducted to reveal the relationship among abundances of MPs with every polymer type, depth, sampling zone, and size range. All the data analyses were performed with SPSS 19.0 (IBM, New York, NY, USA) and Origin 2021 (OriginLab, Northampton, MA, USA). The curves of polymer spectra were visualized by EZ OMNIC. All the figures were drawn in Excel and Origin 2021.

2.6. Error and Accuracy

All experimental steps were strictly conducted according to previous relevant standards to ensure a high accuracy of results, but inevitable errors still existed. For site 5, the GPS coordinates were 34°35.639′ N 135°14.259′ E on 15 September 2021 and 34°35.602′ N 135°14.030′ E on 5 September 2022, respectively. The distance between the two GPS coordinates was about 358 m. Because the berths of experimental ship could not be exactly ensured at the same coordinate of latitude and longitude every sampling time, slight differences of actual geographical location for one site at every sampling time are always inevitable. MPs' abundance at one site among different sampling times may be affected by tidal range and current. In addition, as all MPs on filter could not be counted simultaneously, there must have been some errors in estimating the total MPs' abundance. Moreover, some MPs that had undergone long-term weathering always showed a low peak of FTIR spectra, which made their matching degree less than 70%. These particles always resulted in a "no-polymers" matching and were excluded from the identification of MPs; this may have led to an underestimation of the real MPs abundance.

3. Results and Discussion

3.1. Spatial Distribution of MPs

As shown in Figure 1, MPs in the S-SML and bulk water in Osaka Bay exhibited higher concentrations along navigation routes (S4, S5, S6 and S7) than that in the coastal area (S1, S2 and S3) and at the center of the bay (S8). Especially in the S-SML, MPs' abundance along navigation routes was significantly higher than that in the coastal area (one-way ANOVA Tamhane's T2 post hoc test, $p < 0.05$) and in the center (one-way ANOVA Tamhane's T2 post hoc test, $p < 0.05$).

The average abundance of MPs in the coastal area was 402 $\pm$ 179 items/kg (ranging from 104 to 575 items/kg) in the S-SML, about 10 times higher than the 43 $\pm$ 16.4 items/kg (ranging from 21.8 to 66.6 items/kg) in bulk water. In addition, the average abundance of MPs at S1, S2, and S3 was 457 $\pm$ 7.78, 373 $\pm$ 243, and 390 $\pm$ 255 items/kg in the S-SML and 48.7 $\pm$ 4.67, 43.2 $\pm$ 16.4, and 38.5 $\pm$ 24.5 items/kg in bulk water, respectively. The small difference in MPs' abundance both in the S-SML and in bulk water between S1, S2, and S3 indicated a similar MP pollution distribution in the coastal area.

The average abundance of MPs along navigation routes was 1360 $\pm$ 1050 items/kg in S-SML, about 20 times higher than of the 68.3 $\pm$ 49.3 items/kg in bulk water. The maximum abundance among all sampling sites was found at S6, with 2310 $\pm$ 1390 items/kg in the S-SML and 108 $\pm$ 73.3 items/kg in bulk water. Other sites along navigation routes had similar MP loads. The average abundance of MPs at S4, S5, and S7 was 1070 $\pm$ 1120, 1020 $\pm$ 662, and 1040 $\pm$ 773 items/kg in the S-SML and 56.6 $\pm$ 16.2, 71.2 $\pm$ 44.8, and 43.9 $\pm$ 47.1 items/kg in bulk water, respectively. Most previous studies have shown that MPs' abundance in the coastal area is always higher than that in water areas far from shore [17,34–36]; however, in this study, although navigation routes were far from shore, MPs' abundance therein was significantly larger than that in the coastal area. We believe that MPs newly generated along navigation routes had replaced the ones lost due to long-distance transportation from nearshore to further sea areas; this caused an MP abundance at navigation routes both in the S-SML and bulk water that was about 2–4 times larger than in the coastal area. This result suggests that the number of marine-based MPs generated during navigation far exceed the number originating from land-based MPs, which is quite different from the widely cited 80% terrestrial-to-20% marine-based MP debris ratio for MPs. Similar results to ours were found in a study in the German Bight [19], which also corroborates that these MPs may result from the generation of particles from antifouling and marine paints during the navigating process.

As shown in Figure 1, the average abundance of MPs at the center of the bay (S8) was 261 $\pm$ 240 items/kg in S-SML, about nine times higher than the 33.1 $\pm$ 17 items/kg (ranging from 23.4 to 50.9 items/kg) in bulk water. These numbers were the lowest among all sampling sites, and they also reflected the decrease and loss of MPs from nearshore to offshore.

Figure 1 shows the average abundance of MPs among all different sampling times for every sampling site. However, notably, as shown in Figure S3, the fluctuation in MPs' abundance among different samplings was extraordinarily large in the S-SML both at S4 (ranging from 342 to 1070 items/kg with a CV of 105%) and S8 (ranging from 62.7 items/kg to 528 items/kg with a CV of 92%). This result may be due to the exchange of seawater through the Akashi Strait in the western side of Osaka Bay. Freshwater inflow from all rivers on lands to the Seto Inland Sea has been reported to be approximately 14 km^3/year (38 $\times$ 10^6 m^3/day) [37], 91% of which corresponding to seawater flows through the Akashi Strait from Harimanada [38]. As such, original MPs accumulated at S4 and S8 might be diluted by freshwater flows and then transported to the eastern part of Osaka Bay, which led to the relatively lower MPs' abundances. In total, among all sampling sites, MPs' abundance in the S-SML always showed larger CV values and presented more intense fluctuation than that in bulk water. At the bay center, where all seawater circulation can flow through, the CV value was 110% in the S-SML and 36% in bulk water, respectively. Thus, we can formulate a hypothesis that in the highly mobile and easily diffused S-SML, MPs' abundance constantly changes, and a large fluctuation is noted in the process of long-distance transportation and migration.

3.2. Size Distribution of MPs

MPs were classified in four size categories: 10–53, 53–125, 125–500, and >500 μm, which accounted for 81.2%, 11.1%, 4.45%, and 3.25% in the S-SML, and for 62.2%, 19.8%, 12%, and 6% in bulk water evenly among all sampling sites, respectively. When distinguishing between substrates, as shown in Figure 2, the tiny MPs (10–53 μm) accounted for 72.1%, 84.6%, and 88.5% of all MPs in the coastal area, along the navigation routes, and at the bay center, respectively, in the S-SML. In contrast, the tiny MPs accounted only for 47.4%, 74.4%, and 59.5% for these three zones, respectively, in bulk water. A higher proportion of tiny MPs (10–53 μm) in the S-SML may result from the accelerated fragmentation of larger MPs, because weathering forces including wind, biofouling, and UV irradiation are always stronger and more frequent in the S-SML than in deep water [33]. Thus, larger MPs in the S-SML always undergo a more intensive weathering and break into tiny MPs more rapidly. In addition, MPs less than 100 μm in the water column are easily captured by hydrophobic substances such as grease and microbubbles, which then move upwards [39,40], and this may also lead to an increase in tiny MPs' abundance in the S-SML. It is noteworthy that these exceptionally well floating tiny MPs (10–53 μm) in the S-SML tend to sink once fouled by algae and invertebrates [2], thus they are transported to the subwater and diffuse the pollution. Hence, further studies should focus on mechanisms of vertical transport of MPs that can be affected by both intrinsic MP buoyancy [41] and hydrodynamic characteristics, including stratification, plume front, and turbulence as well [42]. It should be mentioned that the vertical mixing of plastic particles is probably controlled by wind-induced turbulence, and this leads to a preferential removal of heavier and smaller plastic particles from the sea surface, so traditional surface measurements may have severely underestimated the total plastic content [43]. However, it has been verified that MPs with a size of 10, 100, and 1000 μm maintain their maximum concentration in the top layer of <1 m [44]. Therefore, in this study, MPs in bulk water under 1 m from the sea surface may reflect the relatively real MPs' abundance. In addition, even if the S-SML is disrupted by wind turbulence, it can still be rapidly reformed without depletion of organic matter within 1 min [23]. Therefore, the influence of turbulence on MPs' abundance in the S-SML can also be excluded in this study.

As shown in Figure 2b, both in the S-SML and bulk water, as MP sizes increased, their relative abundances decreased (simple linear regression, $r^2 = 0.646$, $p < 0.01$ in the S-SML and $r^2 = 0.604$, $p < 0.01$ in bulk water, respectively), which was mostly consistent with previous research results [11]. MPs with a size >500 μm were significantly fewer than MPs of other sizes (one-way ANOVA Tamhane's T2 post hoc test, $p < 0.05$). MPs with a size of 10–53 μm were the most abundant compared to MPs of other size ranges (one-way ANOVA Tamhane's T2 post hoc test, $p < 0.01$) both in the S-SML and in bulk water. MPs originating from coastal areas always showed high loads in size ranges larger than 53 μm; this may be because these large MPs were newly generated and were not yet weathered and broken into smaller fractions. Among the tiny MPs with a size of 10–53 μm, higher MP loads were mainly along navigation routes rather than in the coastal area and at the bay center. These tiny MPs with larger specific surface areas have increasing sorption capacities, and they can adsorb more pollutants, including persistent organic pollutants (POPs) and endocrine disrupting chemicals, once ingested by organisms, or transported by circulation over long distances. As such, the threats to marine ecosystems of these tiny MPs are multiplied.

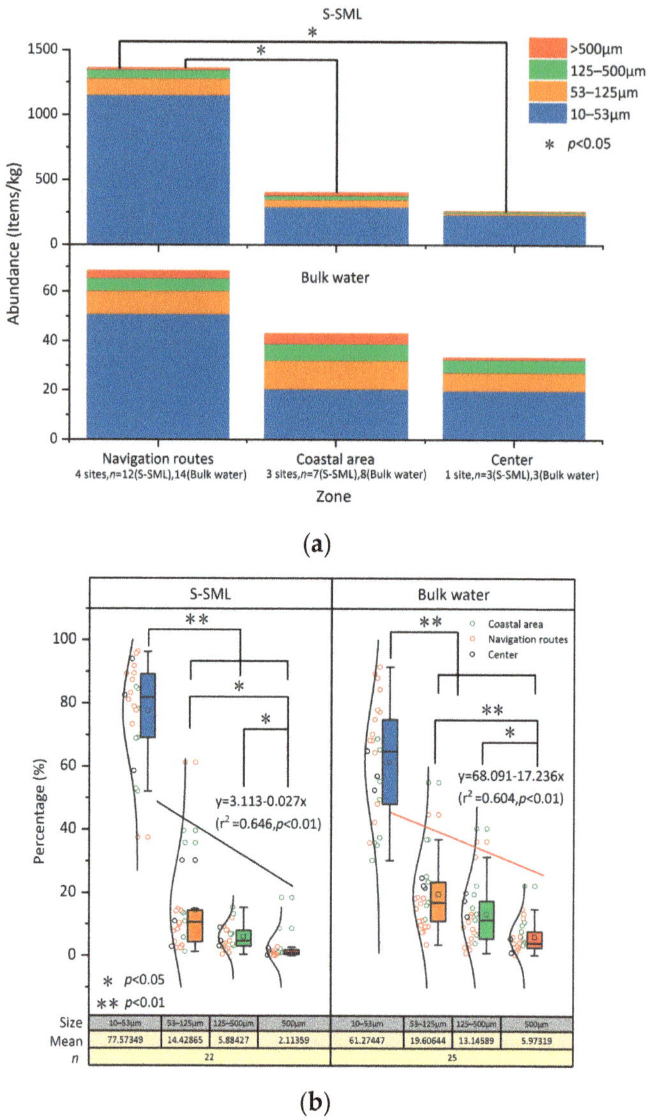

Figure 2. Average abundances of MPs with different size ranges at different sampling zones (**a**) and percentage of MPs with different size ranges at all sampling sites in total (**b**).

3.3. Polymer Composition of MPs

Approximately 11 kinds of polymer types of MPs were detected in Osaka Bay, i.e., acrylic resins mainly represented by polymethyl methacrylate (PMMA), polypropylene (PP), polyethylene (PE), polyester (PET), polystyrene (PS), nylon (PA), polyvinyl chloride (PVC), polyurethane (PU), polycarbonate (PC), ethylene-vinyl acetate copolymer (EVA), and epoxy. The corresponding spectra and pictures observed by FTIR are shown in Figures S4 and S5. Since the abundances of PU, PC EVA, and epoxy were relatively small, these four kinds of polymer types were grouped and defined as "others" in this study. Polymethyl methacrylate (PMMA), the most important type of acrylic polymer, is also called acrylic resin [45]. Besides PMMA, other kinds of acrylic polymers were also detected with relatively less abundance through FTIR, and they were also classified as acrylic resin (PMMA).

As shown in Figure S6, PMMA accounted for the highest proportions of MPs in both the S-SML and bulk water, with 95.1% and 45.6%, respectively, followed by PE, PP, PS, PA, PET, others, and PVC, which accounted for 1.22%, 1.13%, 0.851%, 0.632%, 0.591%, 0.372%, and 0.104% in the S-SML, and 16%, 12.8%, 6.99%, 6.56%, 5.88%, 4.27%, and 1.9% in bulk water, respectively. In addition, PMMA accounted for 96.8% in the S-SML and 49.8% in bulk water among all MPs with a size of 10–53 µm.

As shown in Figure 3, in the S-SML, PMMA dominated all polymer types and accounted for 96.5% (1310 items/kg), 89.9% (362 items/kg), and 88.5% (230 items/kg) along navigation routes, in coastal areas, and at the bay center, respectively. In bulk water, although the percentage of PMMA declined, it was still larger than that of other polymer types and accounted for 49.2% (33.6 items/kg), 36.8% (15.8 items/kg), and 33.7% (11.2 items/kg) at the above sampling zones, respectively. Although the relative density of PMMA is larger than that of the seawater, it still accumulated in the S-SML in large quantities, which can be attributed to the capture capability of the S-SML's tension.

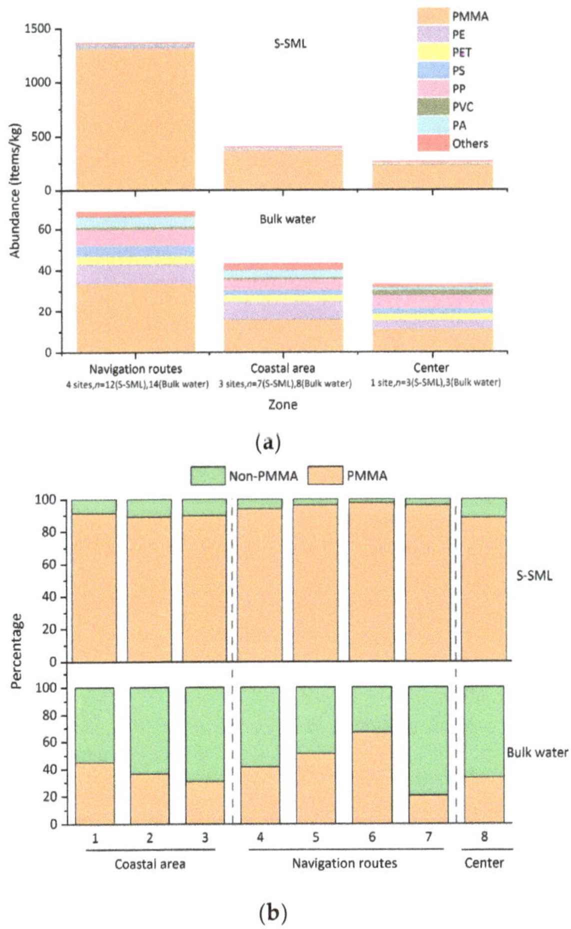

Figure 3. Average abundances of MPs at different sampling zones with different polymer types (**a**) and percentage of MPs with and without PMMA in total (**b**).

These abnormally high abundances of PMMA were different from the results in other relevant studies which revealed that PP and PE were dominant polymer types in aquatic environments [46]. According to the conclusion of a similar study at the German Bight, PMMA predominated in local major shipping lanes and navigation routes [19]. This study also suggested the antifouling and marine paints on ship hulls were plausible sources for this pollution.

3.4. PMMA from Marine and Antifouling Paints

Acrylic polymers such as PMMA stand out among organic coatings because of their nonwettability, chemical inertness, environmental stability, and photostability [47]. Antifouling paint applied on the surface of the hull bottom can prevent the adhesion of fouling organisms and improve speed and reduce fuel consumption for ships. Highly toxic biocides have been incorporated in binders and then are released into water gradually to accomplish antifouling. The use of acrylic resins as binders in antifouling paints can be traced back to 1955 for insoluble matrix or contact paints and to 1974–1985 for self-polishing paints containing tin (TBT-SPC) [48]. Meanwhile, according to the formulas of antifouling paints in Japan, acrylic resins including PMMA were also used as binders [49]. For insoluble matrix or contact antifouling paints, their binders are insoluble in seawater, and pores can be left empty after biocides diffuse out of paint layers, then seawater can spread through these pores and dissolve biocides in deeper binders [48], which may accelerate the abrasion of the honeycomb structure of residual binders with seawater, finally leading to the exfoliation of these acrylic resin fragments without further hydrolysis in a relatively static water column. In addition, for self-polishing antifouling paints that are mainly composed of (meth)acrylic polymer binders [50], as pigments and biocides dissolve faster than the binder hydrolysis, a leached layer is formed at 10–30 μm behind the binder front [51]. Biocides can then be released through a leached layer continuously accompanied by the binder hydrolysis, through which the self-renewing of antifouling paint surface can be achieved [52]. Although (meth)acrylic binders can be soluble in water by hydrolysis of the silyl ester bond [50], for most self-polishing antifouling paints, their antifouling ability mainly depends on the shear of seawater flow and is therefore poor in static seawater [53]. Hence, once the hydrolysis of acrylic binders is prevented by biofilm-covered surfaces, APPs from shipyards by wind and rain runoff may persist in marine environments with a low water flow. In particular, during the navigating process, as hydrolysis occurs, more hydrophilic ester bonds (-COO-) are exposed on the paint surface, which makes the resin on the paint surface brittle and easy to peel off [54]. The friction and collision between the hull and seawater may exacerbate the exfoliation of these resins. Once detached from the hull bottom, these resin fragments can remain in a relatively static water environment without further hydrolysis for a long time, then break into APPs due to UV weathering and biodegradation. By contrast, ships locating in coastal areas are always in a static water environment, and the antifouling paints cannot be hydrolyzed due to the shear of seawater flow; thus, the resins on the surface of paints may not delaminate from the hull bottom. This leads to a lower APPs' abundance in the coastal area compared to navigation routes. Actually, APPs detected in the coastal area mainly come from ship building and boat cleaning at shipyards.

Besides antifouling paints, PMMA is also widely used in marine paints that are applied on the outboard and superstructure of hulls to maintain beauty and gloss. It is noteworthy that acrylic paints are prone to dehiscing and cracking at low temperatures [55], and it has been verified that some acrylic paints become brittle at 5 °C with 50% relative humidity (RH) and even at 13 °C with a lower RH [56]. Therefore, these PMMA marine paints on upper hulls are also prone to becoming brittle and then falling off from the ship structure when navigating, especially in winter, which can result in abundant white PMMA fragments and fiber residues (shown in Figure S4) in the S-SML.

Although MPs in this study were mainly in the shape of fragments, large quantities of PMMA fibers were found in the S-SML at S2, S3, S6, and S7, and their abundances were far larger than that of PMMA fragments. Significantly, most of these PMMA fibers were in the size range of 53–125 μm; therefore, they may have originated from UV weathering and wave fragmentation of MPs with a size of 125–500 μm and >500 μm. If not quickly exchanged with open ocean waters, these PMMA fibers might continuously be broken down and become smaller and more abundant, especially in the local semienclosed waters at S3 and S7. Thus, PMMA fibers may be deemed to be a special shape of disintegrating vectors through which PMMA fragments decompose from larger sizes to smaller sizes.

To further verify whether PMMA and other acrylic polymers in this study derive from antifouling and marine paints, two kinds of paints (as shown in Table S3) with acrylic polymers as binders were chosen to match with MP samples found in Osaka Bay. Paint 1 was used for the outboard and superstructure of hulls to maintain beauty and gloss. As shown in Figure 4, the FTIR spectra of sample 1, which belonged to a PMMA fragment in the S-SML, matched with that of paint 1. For sample 1, a peak at 2942 cm^{-1} showed the stretching vibrations of C-H of methylene, a peak at 1731 cm^{-1} attributed to the bending of saturated C=O-, and the asymmetric stretching vibration of ether C-O-C showed between 1300 and 1000 cm^{-1} and had a peak at 1281 cm^{-1}, according to which the polymer type of sample 1 was identified as PMMA. As the FTIR spectra of sample 1 were clearly presented in all PMMA samples and consistent with that of paint 1, these PMMA samples were assumed to originate from marine paints on the upper structures of ships.

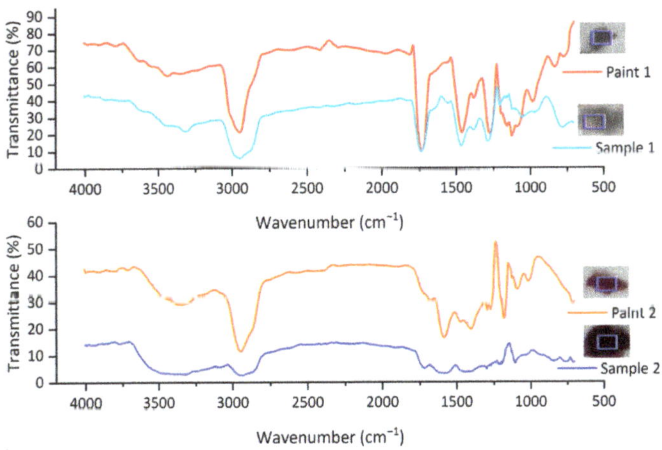

Figure 4. FTIR spectra of polymer types from MPs in Osaka Bay and selected marine and antifouling paints.

Meanwhile, since antifouling paint is one potential source of APPs, the match between MP sample and selected paint 2, one kind of antifouling paint, was conducted to verify this hypothesis. As shown in Figure 4, sample 2, which was one MP fragment in the S-SML, was determined as an acrylic polymer but not as PMMA, and its FTIR spectra showed a significant peak at 2939 cm^{-1} as C-H stretching vibrations, 1714 cm^{-1} as C=O stretching vibrations and 1100, 1204, and 1293 cm^{-1} as C-O-C stretching vibrations, respectively. The FTIR spectra of both sample 2 and paint 2 were consistent, thus pervasive acrylic MPs represented by sample 2 may derive from antifouling paints. Most notably, the main polymer component in acrylic paints is not just confined to PMMA monomer; other acrylic derivative polymers containing acrylate and methacrylate are also widely used in acrylic paints as binders. The spectra peaks of two MP samples in the S-SML were both smaller than those of nonexposed paint samples, which can reflect a higher weathering degree of MP samples that had been floating in seawater for months and years. Therefore, this study is the first to provide empirical evidence of the link between acrylic resins and marine and antifouling paints.

To exclude the possibility that abundant PMMA was generated from the experimental ship, antifouling paints were obtained from the bottom hull of the experimental ship and analyzed by FTIR with a spectral result of polyamide 6:poly(ethylene-co-propylene) 4:1, i.e., a blend of PA 6 and PP-PE with a ratio of 4:1, as shown in Figure S7. Approximate transmittance values at peaks of spectral reflected a similar weathering degree for both MP samples in bulk water and antifouling paints on the experimental ship. As such, some MPs of the polymer type PA 6:PE-PP 4:1 with small abundances could only come from antifouling paint on the experimental ship. Given that PA has been used in the manufacturing of antifouling paint to modify surfaces for reducing bacterial adhesion [57], different from traditional PA fibers from the laundry of textiles and clothes, PA fragments found in this study may also come from marine coatings.

3.5. Difference between PMMA and No-PMMA MPs

As shown in Figure 5 in S-SML, the average abundance of PMMA was 1310 items/kg along navigation routes, 362 items/kg in the coastal area, and 230 items/kg at the bay center, respectively, while 33.6 items/kg, 15.8 items/kg, and 11.2 items/kg were found in corresponding zones, respectively, in bulk water. PMMA MPs with a size of 10–53 mm contributed the most in both S-SML (85.4% along navigation routes, 75% in the coastal area, and 93% at the bay center) and in bulk water (80.1% along the navigation routes, 47.2% in the coastal area, and 77.3% at the bay center). Especially in the S-SML, the abundance of PMMA MPs with a size of 10–53 μm along the navigation routes was significantly higher than that in the coastal area (one-way ANOVA Tamhane's T2 post-hoc test, $p < 0.05$) and at the bay center (one-way ANOVA Tamhane's T2 post hoc test, $p < 0.05$). Also, the PMMA abundances of all size ranges were higher along the navigation routes than in the coastal area (one-way ANOVA Tamhane's T2 post hoc test, $p < 0.05$) and at the bay center (one-way ANOVA Tamhane's T2 post hoc test, $p < 0.05$). However, no significant differences in PMMA abundance were found in bulk water for all sampling zones. The abnormally high PMMA abundance along the navigation routes clearly shows that shipping activities brought about massive acrylic marine and antifouling paint particles and produced large amounts of PMMA MPs with a size of 10–53 μm in the S-SML that may persist along the shipping lanes for months and years.

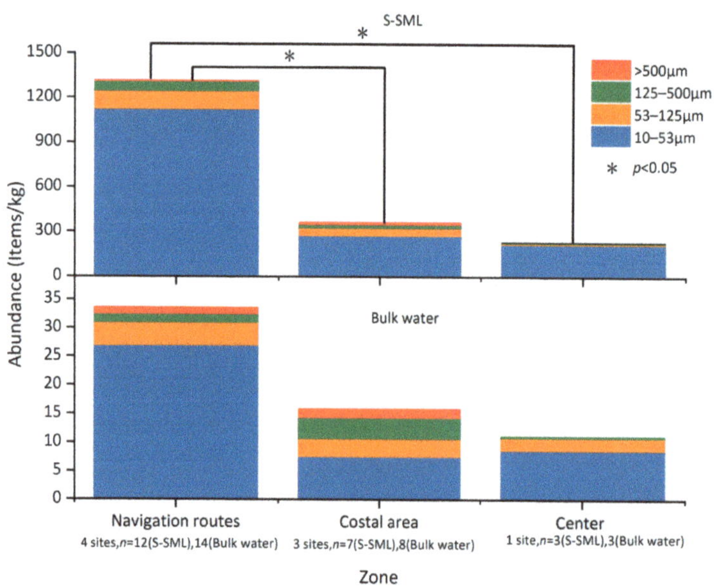

Figure 5. Average abundances of PMMA MPs with different size ranges at different sampling zones.

Although the PMMA abundance in the S-SML was significantly larger than that in bulk water, as shown in Figure 6, MPs that did not belong in the PMMA category (non-PMMA) exhibited a similar abundance level of less than 16 items/kg in both S-SML and bulk water. It could also be verified by the correlation analysis in Figure S8, that PMMA abundance was negatively correlated with water depth, while this was not the case for other polymers. As shown in Figure 5, PMMA abundance was the determining factor for the significant differences between S-SML and bulk water for the total MPs' abundance. It was also a significant factor for the obvious differences in MPs' abundances between the navigation routes, the coastal area, and the bay center. Compared with PMMA, non-PMMA polymers slightly contributed to all polymer types' loads both in S-SML and in bulk water. To some extent, the PMMA abundance can even reflect the whole MP loads in the S-SML. As shown in Figure S9, the result of the correspondence analysis showed that PMMA MPs with a size of 10–53 μm along the navigation routes in the S-SML dominated the whole MP loads in Osaka Bay.

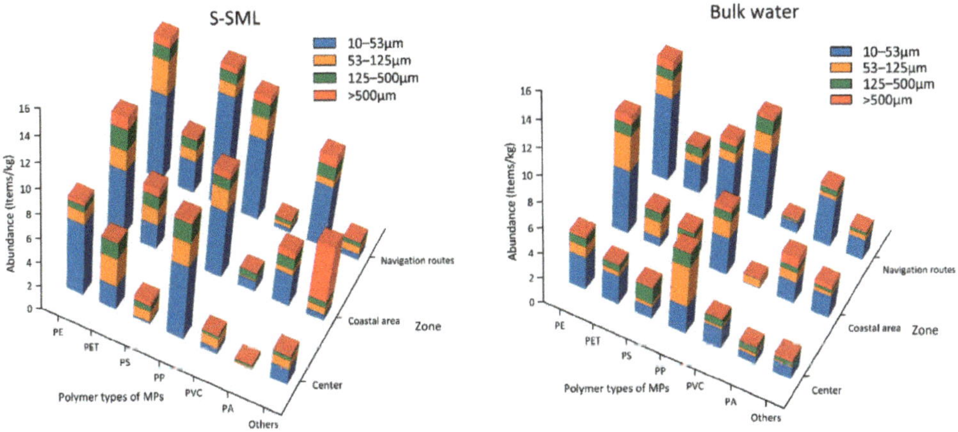

Figure 6. Average abundances of non-PMMA MPs with different size ranges in different sampling zones.

Among non-PMMA polymers, PP and PE presented higher abundances than other polymers both in S-SML and in bulk water. In addition, PET, PS, and PA were also more abundant than PVC in general. As shown in Figure S8, for PMMA, PE, PET, PP, PS, and PVC, their abundances increased as their size became smaller.

3.6. Sources of Polymer Types of MPs

For assessing the contribution of polymer types of MPs on polymer abundances and on sampling sites, a principal component analysis (PCA) was performed, and the loadings are shown in Figure 7a,b for the S-SML and bulk water, respectively. For the S-SML, two principal components explained 63.1% of the variance in the data, where PC1 contributed 36.5% of the total variance and with positive loadings of PMMA, PA, and PE. PC2 contributed 26.6% of the total variance, with positive loadings of PMMA, PP, and PET. For bulk water, two principal components explained 71.4% of the variance in the data, where PC1 contributed 45.4% of the total variance and with positive loadings of PMMA, PP, and PVC. PC2 contributed 26% of the total variance, with positive loadings of PE and PS.

According to relevant previous research, sources of all polymer types were categorized as shown in Figure S10 [58–63]. Except from marine and antifouling paints, PMMA can also originate from terrestrial paints such as those used in building surfaces, road markings, and so on. In addition, PMMA is also the main component of plexiglass, which is widely used in windows, lamps, and dashboards for everyday use, as well as in the manufacturing of airplanes, vehicles, and instruments. Notably, the United States and Japan have made mandatory provisions in their laws that primary and secondary schools and kindergarten buildings must use plexiglass. Furthermore, PMMA microbeads that are used in cosmetics can enter sewage treatment systems directly and can then be released into aquatic environments as primary microplastics. The occurrence of PP and PE is consistent with their huge global production and wide use in modern life, such as plastic bottles, bags, food packing films, fishing gear, caps, and containers due to their low cost and good performance.

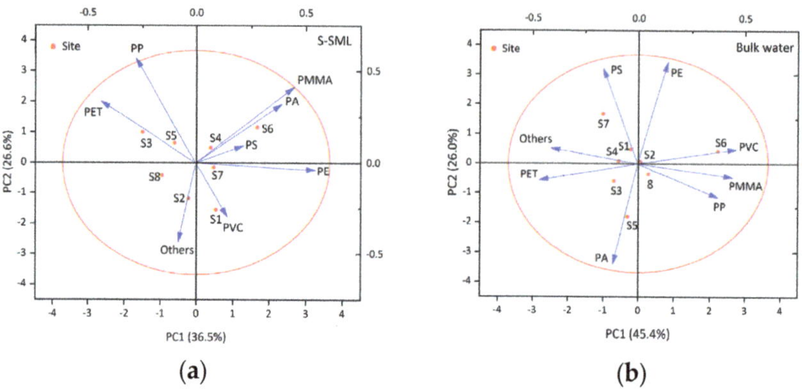

Figure 7. Principal component analysis (PCA) of MPs of all polymer types in S-SML (**a**) and bulk water (**b**).

In the S-SML, the strong positive loadings of PA, PE and PMMA on PC 1 mainly reflected the emissions of marine and antifouling paints from shipping activities. Site 6 was placed in the direction of the PA loading since it had the highest PA average abundance (14.5 items/kg). In addition, site 6 and site 4—located along navigation routes—were placed in the direction of the PMMA loading because they exhibited the highest PMMA average abundance of 2250 items/kg and 1000 items/kg, respectively, among all sampling sites. Sites 4, 6, and 7 belonging to the navigation routes were aligned on PC1, showing that navigation route sites always generated more paint particles. PC2 was characterized by strong positive loadings of PMMA, PP and PET and showed a linkage with discarded fishing gears generated from fishing activities (fishing nets, ropes, and lines, etc.) according to Figure S10. Site 3 and site 5 were aligned on the direction of PET loading because they exhibited the highest average PET abundance of 9.18 items/kg and 5.64 items/kg. The strong negative loadings of PVC and "others" on PC2 may be due to their relatively low abundance, although PVC had the highest average abundance (3.87 items/kg) at site 1 and "others" had the highest average abundance (11.3 items/kg) at site 2; both site 1 and site 2 are located in coastal areas and they had little relationship with fishing activities. To sum up, the PCA in the S-SML seemed to reveal single marine-based MP pollution sources due to marine activities including shipping and fishing activities.

In bulk water, PC1 showed the positive loadings of PMMA, PP, and PVC indicating diverse MP pollution sources, which may come from domestic consumption such as plastic packages and containers, building, and decorative materials such as PVC pipes and other hard plastics. Site 6 on the direction of PVC loading exhibited the highest PVC average abundance of 2.99 items/kg among all sampling sites. PE and PS's positive loadings in PC2 indicated the source of raw materials used in mechanical and electrical manufacturing. PA and PET showed strong negative loadings on PC1 and PC2 because these are fibrous MPs generated from clothes, curtains, and carpets during laundry, which had little relationship with package and building decoration. Site 5 on the direction of PA loading had the highest PA average abundance of 13.1 items/kg among all sampling sites. In total, PCA in bulk water may reflect a complex land-based MP pollution source from daily production and life.

3.7. Summary of Abundances of MPs in Osaka Bay

MPs were widely distributed in both S-SML and bulk water in Osaka Bay. In the S-SML, the average abundance of MPs ranged from 261 $\pm$ 240 items/kg to 2310 $\pm$ 1390 items/kg, with an average abundance of 903 $\pm$ 921 items/kg. In bulk water, the average abundance of MPs ranged from 33.1 $\pm$ 17 items/kg to 108 $\pm$ 77.3 items/kg, with an average abundance of 55.9 $\pm$ 40.4 items/kg. MPs' abundance in the S-SML was about 15 times larger than that in bulk water. These are the first abundance-related data available for Osaka Bay.

As shown in Table 1, if 0.45 μm was taken as the minimum size standard, MPs' abundance in the S-SML in Osaka Bay was much higher than that in Geoje Island [11] and Jinhae Bay in Korea [64] and in Southampton in the UK [65], and MPs' abundance in bulk water in Osaka Bay was much higher than that in Geojie Island in Korea [11] and Hudson River in US [66]. If 44 μm was taken as the minimum size standard, MPs' abundance in the S-SML in Osaka Bay was higher than that in Winyah Bay and Charleston Harbor [67] in the US and was similar to that on the Incheon/Kyeonggi Coast in Korea [33], and MPs' abundance in bulk water was higher than that on the Incheon/Kyeonggi Coast in Korea [33], in the South China Sea [24], in the Terengganu estuary in Malaysia [68], and on the Shore of Rayong in Thailand [69]. Obviously, the abundance of MPs both in the S-SML and bulk water in Osaka Bay was relatively high in relation to worldwide values.

Table 1. Abundances of MPs in surface water in the literature worldwide.

S-SML				Bulk Water			
Size	**Country**	**Site**	**MPs' Abundance (Items/L)**	**Size**	**Country**	**Site**	**MPs' Abundance (Items/L)**
>10 μm	Japan	Osaka Bay (This study)	903	>10 μm	Japan	Osaka Bay (this study)	55.9
>0.75 μm	Korea	Geoje Island ([11])	211	>0.75 μm	Korea	Geoje Island ([11])	0.946
>0.75 μm	Korea	Jinhae Bay ([64])	182	>0.45 μm	US	Hudson River ([66])	0.980
>0.45 μm	UK	Southampton ([65])	75.4	>50 μm	Malaysia	Terengganu estuary ([68])	0.546
>53 μm	Japan	Osaka Bay (This study)	170	>53 μm	Japan	Osaka Bay (this study)	21.1
>50 μm	Korea	Incheon/ Kyeonggi Coastal ([33])	153	>50 μm	Korea	Incheon/ Kyeonggi Coastal ([33])	1.60
>63 μm	US	Winyah Bay ([67])	30.8	>44 μm	China	South China Sea ([24])	2.57
>63 μm	US	Charleston Harbor ([67])	6.60	>75 μm	Thailand	Shore of Rayong ([69])	1.78

4. Conclusions

Our research highlighted the occurrence and distribution of MP pollution in the S-SML in Osaka Bay for the first time. Abundant MPs belonging in the PMMA category were detected both in the S-SML and in bulk water. MPs' abundances in Osaka Bay were relatively higher than those in the marine areas of other countries. MPs' concentrations were higher along navigation routes than in the coastal area and at the bay center, wherein MPs with a size of 10–53 μm were the most abundant category among all size ranges. An FTIR spectra analysis verified an abnormally high abundance of MPs, especially PMMA MPs, along navigation routes, suggesting a pollution source from shipping activities. Although MPs' abundance in the S-SML was significantly larger than that in bulk water, an abundance of MPs with other polymers besides PMMA was similar between the S-SML and bulk water. As such, PMMA was responsible for the large differences in MPs' abundance among navigation routes, coastal waters, and the bay center. It was also responsible for the significant differences in MPs' concentration between the S-SML and bulk water.

Supplementary Materials: The following supporting information can be downloaded at: https://www.mdpi.com/article/10.3390/jox13040044/s1, Table S1: Sampling date, sites, and water temperature; Table S2: Parameters of water samplings; Table S3: Parameter of two kinds of selected marine paint and antifouling paints; Figure S1: Relevant filtering and dealing process of laboratory treatment; Figure S2: Observing method of MPs on filters; Figure S3: Average abundance of MPs at different sampling sites; Figure S4: FTIR spectra of polymer types of MPs; Figure S5: Pictures of polymer types of MPs; Figure S6: Average percentage of every polymer type of MPs both in S-SML and bulk water; Figure S7: Spectra of APPs on experimental ship and MPs in bulk water; Figure S8: Correlation analysis between polymer abundances and seawater and MPs' size factors; Figure S9: Correspondence analysis among depth (S-SML and bulk water), abundance of MPs of all kinds of polymers, sizes, and sampling zones; Figure S10: Source of MPs of all polymer types.

Author Contributions: Conceptualization, H.O.; methodology, M.Z., H.Y. and H.O.; investigation, M.Z., H.Y. and H.O.; data curation, M.Z. and H.Y.; writing—original draft preparation, M.Z., H.Y. and H.O.; writing—review and editing, C.K.Y. and C.E.; supervision, C.E. All authors have read and agreed to the published version of the manuscript.

Funding: This work was supported by the New Energy and Industrial Technology Development Organization, Moonshot Research and Development Program (JPNP 18016-20333325) to H.O.

Institutional Review Board Statement: Not applicable.

Informed Consent Statement: Not applicable.

Data Availability Statement: Additional information is available upon request.

Acknowledgments: We would like to thank Koji HAMASAKI, AORI, The University of Tokyo, for helping us making the RDS-KU1 to collect samples from the S-SML. We also thank Katsumi AOYAMA and Keisuke KITAJIMA, Kobe University Training Ship HAKUO.

Conflicts of Interest: The authors declare no conflict of interest.

References

1. Andrady, A.L. Microplastics in the Marine Environment. *Mar. Pollut. Bull.* **2011**, *62*, 1596–1605. [CrossRef] [PubMed]
2. Barnes, D.K.A.; Galgani, F.; Thompson, R.C.; Barlaz, M. Accumulation and fragmentation of plastic debris in global environments. *Phil. Trans. Soc. B* **2009**, *364*, 1985–1998. [CrossRef] [PubMed]
3. Lebreton, L.; Slat, B.; Ferrari, F.; Sainte-Rose, B.; Aitken, J.; Marthouse, R.; Hajbane, S.; Cunsolo, S.; Schwarz, A.; Levivier, A.; et al. Evidence that the Great Pacifc Garbage Patch is rapidly accumulating plastic. *Sci. Rep.* **2018**, *8*, 4666. [CrossRef] [PubMed]
4. Zitko, V.; Hanlon, M. Another source of pollution by plastics: Skin cleansers with plastic scrubbers. *Mar. Pollut. Bull.* **1991**, *22*, 41–42. [CrossRef]
5. Gregory, M.R. Plastic 'scrubbers' in hand cleansers: A further (and minor) source for marine pollution identified. *Mar. Pollut. Bull.* **1996**, *32*, 867–871. [CrossRef]
6. Browne, M.A.; Galloway, T.; Thompson, R. Microplastic-an emerging contaminant of potential concern? *Integr. Environ. Asses.* **2007**, *3*, 559–561. [CrossRef]
7. Viel, T.; Manfra, L.; Zupo, V.; Libralato, G.; Cocca, M.; Costantini, M. Biodegradation of Plastics Induced by Marine Organisms: Future Perspectives for Bioremediation Approaches. *Polymers* **2023**, *15*, 2673. [CrossRef]
8. Luo, H.W.; Liu, C.Y.; He, D.Q.; Xu, J.; Sun, J.Q.; Li, J.; Pan, X.L. Environmental behaviors of microplastics in aquatic systems: A systematic review on degradation, adsorption, toxicity and biofilm under aging conditions. *J. Hazard. Mater.* **2022**, *423*, 126915. [CrossRef]
9. Yeo, B.G.; Takada, H.; Yamashita, R.; Okazaki, Y.; Uchida, K.; Tokai, T.; Tanaka, K.; Trenholm, N. PCBs and PBDEs in microplastic particles and zooplankton in open water in the Pacifiic Ocean and around the coast of Japan. *Mar. Pollut. Bull.* **2019**, *151*, 110806. [CrossRef]
10. Miller, M.E.; Hamann, M.; Kroon, F.J. Bioaccumulation and biomagnification of microplastics in marine organisms: A review and meta-analysis of current data. *PLoS ONE* **2020**, *15*, 0240792. [CrossRef]
11. Song, Y.K.; Hong, S.H.; Jang, M.; Kang, J.H.; Kwon, O.Y.; Han, G.M.; Shim, W.J. Large Accumulation of Micro-sized Synthetic Polymer Particles in the Sea Surface Microlayer. *Environ. Sci. Technol.* **2014**, *48*, 9014–9021. [CrossRef] [PubMed]

12. Wu, D.; Feng, Y.D.; Wang, R.; Jiang, J.; Guan, Q.Q.; Yang, X.; Wei, H.C.; Xia, Y.K.; Luo, Y.M. Pigment microparticles and microplastics found in human thrombi based on Raman spectral evidence. *J. Adv. Res.* **2023**, *49*, 141–150. [CrossRef] [PubMed]
13. Zhang, N.; Li, Y.B.; He, H.R.; Zhang, J.F.; Ma, G.S. You are what you eat: Microplastics in the feces of young men living in Beijing. *Sci. Total Environ.* **2021**, *767*, 144345. [CrossRef] [PubMed]
14. Van Cauwenberghe, L.; Janssen, C.R. Microplastics in bivalves cultured for human consumption. *Environ. Pollut.* **2014**, *193*, 65–70. [CrossRef] [PubMed]
15. Turner, A. Marine pollution from antifouling paint particles. *Mar. Pollut. Bull.* **2010**, *60*, 159–517. [CrossRef] [PubMed]
16. Turner, A. Paint particles in the marine environment: An overlooked component of microplastics. *Water Res. X* **2021**, *12*, 100–110. [CrossRef]
17. Wang, T.; Zou, X.Q.; Li, B.J.; Yao, Y.L.; Zang, Z.; Li, Y.L.; Yu, W.W.; Wang, W.Z. Preliminary study of the source apportionment and diversity of microplastics: Taking floating microplastics in the South China Sea as an example. *Environ. Pollut.* **2019**, *245*, 965–974. [CrossRef]
18. Wang, T.; Li, B.J.; Yu, W.W.; Zou, X.Q. Microplastic pollution and quantitative source apportionment in the Jiangsu coastal area, China. *Mar. Pollut. Bull.* **2021**, *166*, 112237. [CrossRef]
19. Dibke, C.; Fischer, M.; Scholz-Böttcher, B.M. Microplastic Mass Concentrations and Distribution in German Bight Waters by Pyrolysis−Gas Chromatography−Mass Spectrometry/Thermochemolysis Reveal Potential Impact of Marine Coatings: Do Ships Leave Skid Marks? *Environ. Sci. Technol.* **2021**, *55*, 2285–2295. [CrossRef]
20. Turner, A.; Ostle, C.; Wootton, M. Occurrence and chemical characteristics of microplastic paint flakes in the North Atlantic Ocean. *Sci. Total Environ.* **2022**, *806*, 150375. [CrossRef]
21. Lehmann, M.; Häusl, F.; Gekle, S. Modeling of vertical microplastic transport by rising bubbles. *Microplast. Nanoplast.* **2023**, *3*, 4. [CrossRef]
22. Ye, S.; Andrady, A.L. Fouling of floating plastic debris under Biscayne Bay exposure conditions. *Mar. Pollut. Bull.* **1991**, *22*, 608–613. [CrossRef]
23. Cunliffe, M.; Engel, A.; Frka, S.; Gasparovic, B.; Guitart, C.; Murrell, J.C.; Salter, M.; Stolle, C.; Goddard, R.U.; Wurl, O. Sea surface microlayers: A unified physicochemical and biological perspective of the air−ocean interface. *Prog. Oceangr.* **2013**, *109*, 104–116. [CrossRef]
24. Cai, M.G.; He, H.X.; Liu, M.Y.; Li, S.W.; Tang, G.W.; Wang, W.M.; Huang, P.; Wei, G.; Lin, Y.; Chen, B.; et al. Lost but can't be neglected: Huge quantities of small microplastics hide in the South China Sea. *Sci. Total Environ.* **2018**, *633*, 1206–1216. [CrossRef]
25. Imhof, H.K.; Laforsch, C.; Wiesheu, A.C.; Schmid, J.; Anger, P.M.; Niessner, R.; Ivleva, N.P. Pigments and plastic in limnetic ecosystems: A qualitative and quantitative study on microparticles of different size classes. *Water Res.* **2016**, *98*, 64–74. [CrossRef]
26. Van Cauwenberghe, L.; Devriese, L.; Galgani, F.; Robbens, J.; Janssen, C.R. Microplastics in sediments: A review of techniques, occurrence and effects. *Mar. Environ. Res.* **2015**, *111*, 5–17. [CrossRef]
27. Dris, R.; Imhof, H.; Sanchez, W.; Gasperi, J.; Galgani, F.; Tassin, B.; Laforsch, C. Beyond the ocean: Contamination of freshwater ecosystems with (micro-)plastic particles. *Environ. Chem.* **2015**, *12*, 539–550. [CrossRef]
28. Baldwin, A.K.; Spanjer, A.R.; Rosen, M.R.; Thom, T. Microplastics in lake mead national recreation area, USA: Occurrence and biological uptake. *PLoS ONE* **2020**, *15*, e0228896. [CrossRef]
29. Wang, Z.; Su, B.; Xu, X.; Di, D.; Huang, H.; Mei, K.; Dahlgren, R.A.; Zhang, M.; Shang, X. Preferential accumulation of small (<300 μm) microplastics in the sediments of a coastal plain river network in eastern China. *Water Res.* **2018**, *144*, 393–401.
30. Nishiyama, K.; Yano, T.; Suzuki, H.; Iida, H. Meteorology and Oceanography in the Seto Inland Sea. *Mar. Pollut. Bull.* **1991**, *23*, 5–9. [CrossRef]
31. Chang, P.H.; Guo, X.Y.; Takeoka, H. A numerical study of the seasonal circulation in the Seto Inland Sea, Japan. *J. Oceanogr.* **2009**, *65*, 721–736. [CrossRef]
32. Oo, P.Z.; Boontanon, S.K.; Boontanon, N.; Tanaka, S.; Fujii, S. Horizontal variation of microplastics with tidal fluctuation in the Chao Phraya River Estuary, Thailand. *Mar. Pollut. Bull.* **2021**, *173*, 112933. [CrossRef] [PubMed]
33. Chae, D.H.; Kim, I.S.; Kim, S.K.; Song, Y.K.; Shim, W.J. Abundance and Distribution Characteristics of Microplastics in Surface Seawaters of the Incheon/Kyeonggi Coastal Region. *Arch. Environ. Contam. Toxicol.* **2015**, *69*, 267–278. [CrossRef]
34. Pan, Z.; Guo, H.G.; Chen, H.Z.; Wang, S.M.; Sun, X.W.; Zou, Q.P.; Zhang, Y.B.; Lin, H.; Cai, S.Z.; Huang, J. Microplastics in the Northwestern Pacific: Abundance, distribution, and characteristics. *Sci. Total Environ.* **2019**, *650*, 1913–1922. [CrossRef] [PubMed]
35. Cincinelli, A.; Scopetani, C.; Chelazzi, D.; Lombardini, E.; Martellini, T.; Katsoyiannis, A.; Fossi, M.C.; Corsolini, S. Microplastic in the surface waters of the Ross Sea (Antarctica): Occurrence, distribution and characterization by FTIR. *Chemosphere* **2017**, *175*, 391–400. [CrossRef]
36. D'Hont, A.; Gittenberge, A.; Leuven, R.S.E.W.; Hendriks, A.J. Dropping the microbead: Source and sink related microplastic distribution in the Black Sea and Caspian Sea basins. *Mar. Pollut. Bull.* **2021**, *173*, 112982. [CrossRef]
37. Yamamoto, T.; Kitamaru, T.; Matsuda, O. River inputs of fresh water, total nitrogen and total phosphorus into the Seto Island Sea. *J. Fac. Appl. Biol. Sci. Hiroshima Univ.* **1996**, *35*, 81–104.
38. Hoshika, A.; Tanimoto, T.; Mishima, Y. Current and material transport at Tomogashima Strait, Japan. *J. Oceanogr.* **1999**, *55*, 427–437. [CrossRef]
39. So, W.K.; Ghan, K.; Not, C. Abundance of plastic microbeads in Hong Kong coastal water. *Mar. Pollut. Bull.* **2018**, *133*, 500–505. [CrossRef]

40. Kalčíková, G.; Alič, B.; Skalar, T.; Bundschuh, M.; Gotvajn, A.Ž. Wastewater treatment plant effluents as source of cosmetic polyethylene microbeads to freshwater. *Chemosphere* **2017**, *188*, 25–31. [CrossRef]
41. Ryan, P.G. Does size and buoyancy affect the long-distance transport of floating debris? *Res. Res. Lett.* **2015**, *10*, 084019. [CrossRef]
42. Krelling, A.P.; Souza, M.M.; Williams, A.T.; Turra, A. Transboundary movement of marine litter in an estuarine gradient: Evaluating sources and sinks using hydrodynamic modelling and ground truthing estimates. *Mar. Pollut. Bull.* **2017**, *119*, 48–63. [CrossRef] [PubMed]
43. Zhang, H. Transport of microplastics in coastal seas. *Estuar. Coast. Shelf Sci.* **2017**, *199*, 74–86. [CrossRef]
44. Enders, K.; Lenz, R.; Stedmon, C.A.; Nielsen, T.G. Abundance, size and polymer composition of marine microplastics $\geq$10 μm in the Atlantic Ocean and their modelled vertical distribution. *Mar. Pollut. Bull.* **2015**, *100*, 70–81. [CrossRef]
45. Raita, G.; Takashi, I. Poly(methyl methacrylate) (PMMA). In *Encyclopedia of Polymeric Nanomaterials*; Kobayashi, S., Müllen, K., Eds.; Springer: Berlin/Heidelberg, Germany, 2015.
46. Madhuraj, P.K.; Appu, J.B.; Amal, R.; Riya, K.A.; Abhinab, B.; Suja, P.D. Machine learning aided meta-analysis of microplastic polymer composition in global marine environment. *J. Hazard. Mat.* **2022**, *440*, 129801.
47. González-Reyna, M.A.; Espinosa-Medina, M.A.; Rodrigo Esparza, A.R.; Hernández-Martinez, A.R.; Maya-Cornejo, J.; Estévez, M. Anticorrosive Effect of the Size of Silica Nanoparticles on PMMA-Based Hybrid Coatings. *J. Mater. Eng. Perform.* **2021**, *30*, 1054–1065. [CrossRef]
48. Almeida, E.; Diamantino, T.C.; Soua, O. Marine paints: The particular case of antifouling paints. *Prog. Org. Coat.* **2007**, *59*, 2–20. [CrossRef]
49. Tsuboi, M.; Yoshikawa, E.; Arimura, H.; Kozono, S.; Nakamura, N. Silyl (Meth)acrylate Copolymer-Based Antifouling Marine Coating Composition. Japan Patent JP2001081147, 27 March 2001. (In Japanese)
50. Lejars, M.; Margaillan, A.; Bressy, C. Siloxy Silylester Methacrylate Diblock Copolymer-Based Coatings with Tunable Erosion and Marine Antifouling Properties. *Appl. Polym. Mater.* **2020**, *2*, 3291–3300. [CrossRef]
51. Howell, D.; Behrends, B. A methodology for evaluating biocide release rate, surface roughness and leach layer formation in a TBT-free, self-polishing antifouling coating. *Biofouling* **2006**, *22*, 303–315. [CrossRef]
52. Kii, S.; Dam-Johansen, K.; Weinell, C.E.; Pedersen, M.S. Seawater-Soluble Pigments and Their Potential Use in Self-Polishing Antifouling Paints: Simulation-Based Screening Tool. *Prog. Org. Coat.* **2002**, *45*, 423–434.
53. Sha, J.; Chen, R.; Yu, J.; Liu, Q.; Liu, J.; Zhu, J.; Liu, P.; Li, R.; Wang, J. Dynamic multi-level microstructured antifouling surfaces by combining quaternary ammonium modified GO with self-polishing copolymers. *Carbon* **2023**, *201*, 1038–1947. [CrossRef]
54. Sha, J.; Liu, X.; Chen, R.; Yu, J.; Liu, Q.; Liu, J.; Zhu, J.; Liu, P.; Li, R.; Wang, J. Surface hydrolysis-anchored eugenol self-polishing marine antifouling coating. *J. Colloid Interface Sci.* **2023**, *637*, 67–75. [CrossRef] [PubMed]
55. Hayes, J.; Golden, M.; Smith, G.D. From Formulation to Finished Product: Causes and Potential Cures for Conservation Concerns in Acrylic Emulsion Paints. In *Modern Paints Uncovered*, 2nd ed.; Leamer, T., Smithen, P., Kruger, J.W., Schilling, M.R., Eds.; Getty Conservation Institute: Los Angeles, CA, USA, 2007; pp. 58–65.
56. Erlebacher, J.D.; Brown, E.; Mecklenburg, M.F.; Tumosa, C.S. The Effects of Temperature and Relative Humidity on the Mechanical Properties of Modern Painting Materials. *MRS Online Proc. Libr.* **1992**, *267*, 359–370. [CrossRef]
57. Mansourpanah, Y.; Habili, E.M. Preparation and modification of thin film PA membranes with improved antifouling property using acrylic acid and UV irradiation. *J. Membr. Sci.* **2013**, *430*, 158–166. [CrossRef]
58. Geyer, R.; Jambeck, J.R.; Law, K.L. Production, use, and fate of all plastics ever made. *Adv. Sci.* **2017**, *3*, 1700782. [CrossRef] [PubMed]
59. Kelly, M.R.; Lant, N.J.; Kurr, M.; Burgess, J.G. Importance of water-volume on the release of microplastic fibers from laundry. *Environ. Sci. Technol.* **2019**, *53*, 11735–11744. [CrossRef] [PubMed]
60. Park, C.H.; Kang, Y.K.; Seung-Soon, I. Biodegradability of cellulose fabrics. *J. Appl. Polym.* **2004**, *94*, 248–253. [CrossRef]
61. Siracusa, V.; Rocculi, P.; Romani, S.; Dalla Rosa, M. Biodegradable polymers for food packaging: A review. *Trends. Food. Sci. Technol.* **2008**, *19*, 634–643. [CrossRef]
62. Yu, D.; Peng, X.B.; Wang, J.P.; Wang, J.D.; Bao, K. Occurrence of microplastics in the beach sand of the Chinese inner sea: The Bohai Sea. *Environ. Pollut.* **2016**, *214*, 722–730. [CrossRef]
63. Zhao, D.; Ran, J.M.; Teng, W.; Liu, J.; Liu, Y.L.; Yin, X.N. Microplastic pollution in sediments from the Bohai Sea and the Yellow Sea, China. *Sci. Total Environ.* **2018**, *640–641*, 637–645. [CrossRef]
64. Song, Y.K.; Hong, S.H.; Jang, M.; Han, G.M.; Shim, W.G. Occurrence and Distribution of Microplastics in the Sea Surface Microlayer in Jinhae Bay, South Korea. *Arch. Environ. Contam. Toxicol.* **2015**, *69*, 279–287. [CrossRef] [PubMed]
65. Stead, J.L.; Cundy, A.B.; Hudson, M.D.; Thompson, C.E.L.; Williams, I.D.; Russell, A.E.; Pabortsava, K. Identification of tidal trapping of microplastics in a temperate salt marsh system using sea surface microlayer sampling. *Sci. Rep.* **2020**, *10*, 14147. [CrossRef] [PubMed]
66. Miller, R.Z.; Watts, A.J.; Winslow, B.O.; Galloway, T.S.; Barrows, A.P. Mountains to the sea: River study of plastic and non-plastic microfiber pollution in the northeast USA. *Mar. Pollut. Bull.* **2017**, *124*, 245–251. [CrossRef] [PubMed]
67. Gray, A.D.; Werta, H.; Leads, R.R.; Weinstein, J.E. Microplastic in two South Carolina Estuaries: Occurrence, distribution, and composition. *Mar. Pollut. Bull.* **2018**, *128*, 223–233. [CrossRef]

68. Taha, Z.D.; Amin, R.M.; Anuar, S.T.; Nasser, A.A.A.; Sohaimi, E.S. Microplastics in seawater and zooplankton: A case study from Terengganu estuary and offshore waters, Malaysia. *Sci. Total Environ.* **2021**, *786*, 147466. [CrossRef] [PubMed]
69. Prarat, P.; Hongsawat, P. Microplastic pollution in surface seawater and beach sand from the shore of Rayong province, Thailand: Distribution, characterization, and ecological risk assessment. *Mar. Pollut. Bull.* **2022**, *174*, 113–200. [CrossRef]

Journal of *Xenobiotics*

MDPI

Article

Pyriproxyfen Contamination in *Daphnia magna*: Identifying Early Warning Biomarkers

Beatriz Salesa [1,*], Javier Torres-Gavilá [2], María Dolores Ferrando-Rodrigo [3] and Encarnación Sancho [3]

1 Biomaterials and Bioengineering Lab, Centro de Investigación Traslacional San Alberto Magno, Universidad Católica de Valencia San Vicente Mártir, 46001 Valencia, Spain

2 Instituto de Investigación en Medio Ambiente y Ciencia Marina (IMEDMAR-UCV), c/Guillem de Castro 94, 46001 Valencia, Spain; javier.torres@ucv.es

3 Laboratory of Ecotoxicology, Functional Biology and Physical Anthropology Department, Faculty of Biology, University of Valencia, Dr. Moliner 50, 46100 Valencia, Spain; dolores.ferrando@uv.es (M.D.F.-R.); encarna.sancho@uv.es (E.S.)

* Correspondence: beatriz.salesa@ucv.es

Abstract: Pyriproxyfen is an insecticide currently employed in numerous countries for the management of agricultural and indoor pests. Several studies indicate that this insecticide has been detected in multiple rivers, with concentrations reaching as high as 99.59 ng/L in the Júcar River in Spain. Therefore, the determination of some biochemical and genetic effects of this insecticide on aquatic organisms could serve as an early warning mechanism to identify potential disruptions in various biomarkers. Based on this, *Daphnia magna* organisms were exposed to pyriproxyfen sublethal concentrations for 21 days. Some biochemical parameters, including cholesterol, triglycerides, glucose, lactate, and LDH activity, were determined. Additionally, some genetic biomarkers associated with oxidative stress, heat shock proteins, lipid metabolism, hemoglobin, metallothioneins, and vitellogenin synthesis were evaluated in daphnids exposed to the insecticide for 21 days. LDH activity increased significantly in those daphnids exposed to the highest insecticide concentration (14.02 μg/L), while cholesterol levels decreased significantly. In contrast, glucose, total proteins, and triglycerides remained unaffected in *D. magna* exposed to pyriproxyfen. On the other hand, exposure to the insecticide led to notable alterations in gene expression among individuals. Specifically, genes associated with lipid metabolism and reproduction exhibited a significant reduction in gene expression. *Fabd* expression was decreased by approximately 20% in exposed daphnids, while *vtg* expression was suppressed as much as 80% when compared to control values. Furthermore, it was observed that the *hgb1* and *hgb2* genes, associated with hemoglobin synthesis, exhibited significant overexpression. Notably, the dysfunction observed in both hemoglobin genes was linked to an increase in pigmentation in *Daphnia magna* during the course of the experiment. These alterations in gene expression could serve as effective indicators of early contamination even at low pesticide concentrations.

Keywords: pyriproxyfen; *Daphnia magna*; early biomarkers; gene expression; biochemicals

1. Introduction

Pyriproxyfen, chemically known as 2-[1-methyl-2-(4-phenoxyphenoxy) ethoxy] pyridine, is a powerful insecticide functioning as an insect growth regulator, and it has a broad range of applications for pest control. It is extensively employed in various public health programs, serving as an alternative to organophosphate and pyrethroid pesticides [1]. In fact, it is one of the most efficacious juvenile hormone analogs [2] due to its development and larvae maturation inhibition effect [3], use as a juvenile hormone III mimic [4], in preventing insect hatching success [5], disrupting larval development [6], or inducing morphological abnormalities in adult insects, which decreases their reproductive capacity [7], among others. The effectiveness of its active ingredient was assessed to determine its safety for use as an insecticide in drinking water [8], ensuring minimal adverse effects on non-target aquatic insects and the surrounding environment [9–12].

Pyriproxyfen, known for its relative environmental stability and lipophilic properties (octanol/water partitioning coefficient of Kow 5.6), may have a significant impact in freshwater environments which are close to agricultural areas [13]. Concentrations of this insecticide between 37.74 ng/L and 99.59 ng/L had been detected in western Mediterranean European rivers, such as the Ebro [14] or Júcar [15], derived from agriculture uses.

According to the World Health Organization (WHO) [8], safe concentration levels for this insecticide are set at 10 μg/L for drinking water sources and 100 mg/L for bodies of water. Furthermore, various studies have shown that the effective concentrations utilized for controlling mosquito pests can result in detrimental impacts on a range of terrestrial invertebrates like honeybees, as well as aquatic invertebrates such as amphipods, crabs, and daphnids, affecting both embryos and adult individuals [16–20].

Daphnia magna hold a crucial role as a keystone species in aquatic ecosystems, serving as primary consumers of algae and providing a vital food source for larger invertebrates and fish [21]. Disruptions in the population of primary consumers can contribute to environmental issues (i.e., algal blooms, reduced fish populations, and overall disruption of the aquatic ecosystem).

Invertebrates' early exposure to an endocrine disruptor may result in long-term effects in their endocrine systems, potentially leading to reproductive effects in adult individuals [22,23]. Reproduction in parthenogenetically populations such as *D. magna* could be seriously affected by pyriproxyfen exposure, resulting in an elevated male production [24]. Olmstead and LeBlanc [25] determined pyriproxyfen's male production EC_{50} in *D. magna*, about 170–310 pM (55–100 ng/L). Elevated male production in Daphnia species leads to a decrease in the production of parthenogenic females, a decrease in the population growth rate and, ultimately, in the overall population abundance [25].

Taking into account pyriproxyfen residues resulting from agricultural applications in water during the initial weeks post-application, these levels could potentially reach concentrations that negatively impact invertebrate populations and other primary consumers residing in rice fields, floodwaters, freshwater ponds, and similar environments. Biochemical parameters encompass a wide range of responses, and they can be highly sensitive to various pollutants. The selection of key general metabolites associated with Daphnia's energy metabolism, such as proteins, lipids, glucose, and the lactate pathways, is a straightforward approach, yielding a rapid response for assessment [26]. These biomolecules are crucial for the proper development of an individual, and deviations in their concentrations could be attributed to stress induced by the presence of a contaminating agent [27]. Moreover, an examination of genes associated with various processes, including xenobiotic detoxification pathways (such as metallothionein synthesis), responses to stressors (heat shock protein synthesis), oxidative stress management (involving glutathione-S-transferase and catalase), reproduction (vitellogenin), the regulation of fatty acids (acyltransporter protein synthesis), and hemoglobin production, could serve as an early warning system for assessing the health status of individuals and, ultimately, entire populations.

Rider et al. [28] established that the accumulation of hemoglobin in daphnids is a significant adaptive response that is governed by at least two separate molecular pathways: one driven by endocrine influences activated by terpenoid hormones and another associated with oxygen sensing, which involves the hypoxia-inducible factor. Research into environmental signals capable of triggering neuro-endocrine cascades that govern various physiological processes has revealed that the accumulation of hemoglobin and the production of male offspring are concurrently influenced by the crustacean terpenoid hormone methyl farnesoate, as well as several synthetic analogues like pyriproxyfen. While sublethal effects do not lead to immediate mortality, they have adverse effects on the individual. The physiological changes observed in response to environmental pesticide levels, particularly in invertebrates other than insects, not only offer insights into the ecotoxicological implications of pollutants at biological levels but can also serve as a model for understanding the aquatic toxicity of vertebrates.

The primary goal of this study was to investigate the physiological responses of *Daphnia magna* individuals when exposed to chronic exposure to the insecticide pyriproxyfen, a potent endocrine disruptor in insects. This research was conducted under controlled laboratory settings. Biochemical parameters associated with intermediate metabolism were utilized as critical indicators of insecticide toxicity at the individual level. Additionally, the study explored the expression of certain genes linked to oxidative stress responses (*gst* and *cat* genes), lipid metabolism (*fabd* gene), hemoglobin (*hb1* and *hb2* genes), vitellogenin (*vtg 1* and *vtg 2* genes), heat-shock protein genes (*hsp70* and *hsp 90*), and metallothioneins (*mt-a, mt-b, mt-c* genes), which were considered as potential tools for the early detection of individual dysfunction. The parameters selected could prove to be valuable as a diagnostic test for assessing the well-being of aquatic invertebrates following pesticide spray operations.

2. Materials and Methods

2.1. *Daphnia magna* Culture

Daphnia magna clon K6 [29] were maintained in our laboratory following OECD [30] recommendations for more than 20 years. The culture was set in a controlled climate chamber with a room temperature of 22 ± 1 °C, and a photoperiod of 16:8 h light/dark. Daphnids were kept in well-oxygenated tap water with a pH of 7.9 ± 0.2, alkalinity of 4.1 mmol/L, total hardness of 181.8 ± 18.8 mg/L as $CaCO_3$, and oxygen saturation of more than 90%. The population density was maintained below 40 animals per liter. *Nannochloropsis oculata* algae at a concentration of 5×10^5 cells/mL was used as food for daphnids and were cultivated in our laboratory using Bold's Basal Medium [31] under constant illumination [32,33]. The daphnids used in the experiments were only females, as they reproduced by parthenogenesis in good environmental conditions [33,34].

2.2. Test Chemical and Solutions

Technical product pyriproxyfen (CAS number 95737-68-1, >97% purity), chemical name: 4-phenoxyphenyl (RS)-2-(2-pyridyloxy) propyl ether was provided by The Plant Protection Company S.L (Proplan, Madrid, Spain).

Pyriproxyfen experiments were conducted in our laboratory using five different test concentrations: 3.14, 4.74, 6.12, 8.40, and 14.02 μg/L. These concentrations corresponded to 1/107, 1/71, 1/55, 1/40, and 1/24 of the EC_{50}—48 h value [35]. Ethanol was used as the pyriproxyfen solvent with a maximum concentration of 0.28 μL/L. Previous experiments have confirmed that this ethanol concentration is safe for *Daphnia magna*, and no adverse effects were observed in this species [36]. The pyriproxyfen stock solutions were freshly prepared daily using a renewal medium.

2.3. Experimental Design

In the experimental setup, an individual adult female was randomly selected from the general culture. All her offspring from a single brood were then placed in a separate culture, where they were raised under the same environmental conditions as the general culture. Starting from the third brood onward, the neonates were utilized in the experiments.

For each experiment, 40 neonates (≤24 h old) were randomly distributed into individual 3 L glass beakers. The experimental medium and food (*N. oculata* at a concentration of 5×10^5 cells/mL) were consistently maintained by transferring the *Daphnia* to fresh test solutions or control water every day. The experimental water and food were the same as those used in the general *Daphnia* culture, as previously described, and the experimental maintenance conditions in the climate chamber were in accordance with the conditions mentioned in Section 2.1 (*Daphnia magna* culture). These experiments were conducted for 21 days using five different concentrations of pyriproxyfen, in addition to a control group, all under these specified conditions.

Upon reaching reproductive maturity, the offspring were regularly extracted from the containers during the daily medium renewal to ensure that only adult individuals remained in each culture. After 21 days of exposure to these conditions, each adult *Daphnia* was carefully transferred to a labeled 1.5 mL Eppendorf tube. In order to guarantee complete dryness before preservation, any residual medium was removed. Subsequently, the labeled tubes containing the Daphnia specimens were promptly frozen using liquid nitrogen and stored at a temperature of −80 °C until further analysis.

2.4. Biochemical Analysis

For each pesticide concentration plus control group, 6 replicates (2 individuals/replicate) were performed. Test organisms were placed in microcentrifuge tubes with 30 µL of phosphate buffer (pH 7.2) and homogenized on ice. Subsequently, the homogenate was centrifuged (6 min, 1400× *g*) and the supernatant transferred to a clean tube preserved on ice for its immediate use. All reactions took place at laboratory temperature (25 °C).

Bradford Protein Assay reagent (Bio-Rad, Hercules, CA, USA) according to Bradford (1976) was used to determine the protein content in the samples using bovine serum albumin (Sigma-Aldrich, Zwijndrecht, Switzerland) as standard. Total cholesterol, triglycerides, glucose, and lactate dehydrogenase (LDH) were determined by using commercial Spinreact's kits (refs. 001090, 1001310, 1001200, and 1001260, respectively) following the manufacturer's protocol.

All the assays were adjusted for measurement in microplates. The contents of each sample were determined spectrophotometrically using a Varioskan microplate reader (ThermoScientific, Waltham, MA, USA), at their respective absorbance wavelengths. The data obtained were subsequently normalized to protein content and expressed for each individual *Daphnia*.

2.5. RNA Isolation and Quantitative RT-PCR

To analyze the gene expression, a total of 6 replicates were performed. In each replicate (1 individual per replicate), manual homogenization was carried out using the buffer provided by the RNA purification kit (Norgen, Thorold, ON, Canada). The extraction process was conducted in accordance with the manufacturer's protocol. The concentrations and quality of the extracted RNA were assessed using a Nanodrop™ One instrument (ThermoScientific™, Waltham, MA, USA). Finally, 1 µg of RNA was used by means of the PrimeScript™ RT (Perfect Real Time) kit, reverse transcribing for the cDNA synthesis (Takara Bio Inc., Otsu, Japan).

To assess the expression of the genes selected, RT-PCR was conducted using TB Green Premix Ex Taq (Takara Bio Inc., Otsu, Japan) as the reagent in the 384 QuantStudio 5 real-time PCR detection system (ThermoScientific™, Dreieich, Germany). The thermocycling profile consisted of an initial denaturation step at 95 °C for 10 min, followed by 40 amplification cycles at 95 °C for 15 s, 60 °C for 30 s, and 72 °C for 15 s. GAPDH (glyceraldehyde-3-phospate dehydrogenase) was used as the reference gene. A melting curve analysis was carried out to confirm the unique amplification of the selected genes.

The primer sequences employed in the study are presented in Table 1. We used QuantStudio™ Design and Analysis Software 1.5.1 (ThermoFisher, Waltham, MA, USA) to analyze the data obtained from the real-time PCR analysis.

2.6. Statistical Analysis

GraphPad Prism 7 software (GraphPad Software, Solana Beach, CA, USA) was used for statistical analyses. Kolmogorov–Smirnov and the Levene tests were used in a priory test. Significant differences between groups were assessed using one-way ANOVA, followed by Dunnett's multiple tests as a post hoc analysis. The significance level was predetermined at p-values < 0.05.

Table 1. Primer sequences employed for gene-specific real-time RT-PCR.

Gene	Forward	Reverse	Reference
GAPDH	TGCTGATGCCCCAATGTTTGTTGT	GCAGTTATGGCGTGGACGGTTGT	[37]
vtg1	CCAGCGAATCCTACACCGTCAAG	GAGCCGCACAGACCACAGAG	[38]
vtg2	CGTCCGCCACTGGTTGGGTC	GGGGCAGCCAAGACAGAGCG	
mt-a	GAGCGCCATGCCAAAATCCC	TCGTCGTTGTAAAATCCGGCT	
mt-b	TGGAACCGAATGCAAATGCG	CGGACTTGCATGGACAACTG	
mt-c	AAAGTGTGCCCTCGTTGTCA	CTTACAGTCGTCCCCACACG	
gst	TCAGGCTGGTGTTGAGTTTG	GAGCAAGCATTTGTGCATCA	
cat	TGGCGGAGAAAGCGGTTCAGC	GTGCGTGGTCTCTGGGCGAA	
hsp70	CGACGGCGGGAGATACGCAC	CCACGGAAAAGGTCGGCGCA	
hsp90	CCCTCTGTGACACTGGTATTGGCA	GCCCATGGGTTCTCCATGGTCAG	
fabd	GCCAACTACCTGTATCCTGAATG	GTGGAACGCTCCGCTAACT	
hb1	ACAAATTGCTCTGGTTGCCG	AAGGTTTTTGAGTGCCACGT	[39]
hb2	TGTTACCACCAGTGTCACCA	TATTCAGGGTGGGCCTTGAT	

3. Results and Discussion

After 21 days of *Daphnia magna* chronic pyriproxyfen exposure, no mortality was observed, and the survival rate remained at 100% in both the control groups and the exposed daphnids.

Lactate dehydrogenase activity (LDH), as well as cholesterol levels in pyriproxyfen-exposed *D. magna*, are shown in Figure 1. As we can see in Figure 1A, LDH activity increased significantly ($p < 0.05$) in those daphnids exposed to the highest pesticide concentration (14.02 µg/L), while cholesterol levels decreased significantly ($p < 0.05$) (Figure 1B).

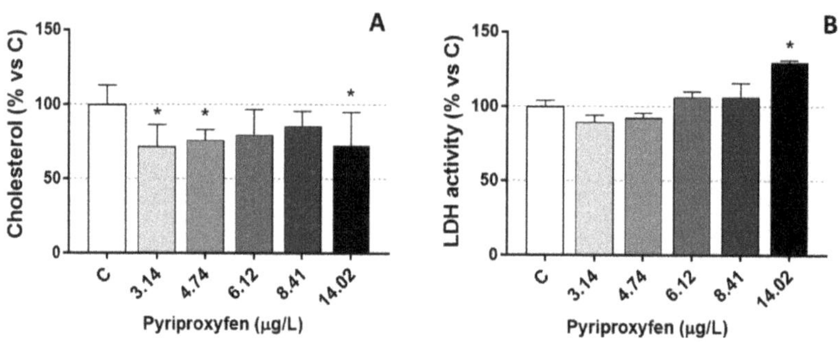

Figure 1. LDH activity (**A**) and cholesterol levels (**B**) in *D. magna* exposed to pyriproxyfen for 21 days, compared (%) to control group C. Results are mean ± SEM (* $p < 0.05$).

In contrast, glucose, total proteins, and triglycerides (Figure 2) remained unaffected in those daphnids exposed to pyriproxyfen. Therefore, the results suggest that the energy reserves of *D. magna* were not significantly affected by the selected pyriproxyfen concentrations during the 21-day period, and the findings do not indicate a substantial toxic stress response.

The increase in LDH activity observed at the highest exposure concentration may be linked to increased resource demand. Wu and Lam [40] previously reported a negative correlation between LDH activity and ambient oxygen levels in *D. magna*, suggesting a potential biochemical adjustment in response to hypoxic conditions. LDH is a cytoplasmic enzyme found in numerous tissues, and its elevation, following intoxication, could signify both damage and inadequate oxygen intake, as well as an additional energy requirement to meet the heightened energy demands of stressed animals. In fact, a transient increase in LDH activity could indicate the induction of anaerobic glycolysis to fulfill the heightened energy demands. In vertebrates like fish, increased anaerobic metabolism is a rapid and clear response to the energy depletion caused by oxygen deprivation [41]. A reduction in lipid reserves is typically associated with an increase in energy demand in pesticide-exposed animals, and lipid catabolism might be an outcome of a toxicant-induced stress situation. However, in the present study we did not find a mobilization of lipid reserves (as triglycerides) and only cholesterol levels decreased in our exposed daphnids. Decreases in both cholesterol and triglycerides were observed in *D. magna* because of chronic exposure to other pesticides, such as the fungicide tebuconazole [42].

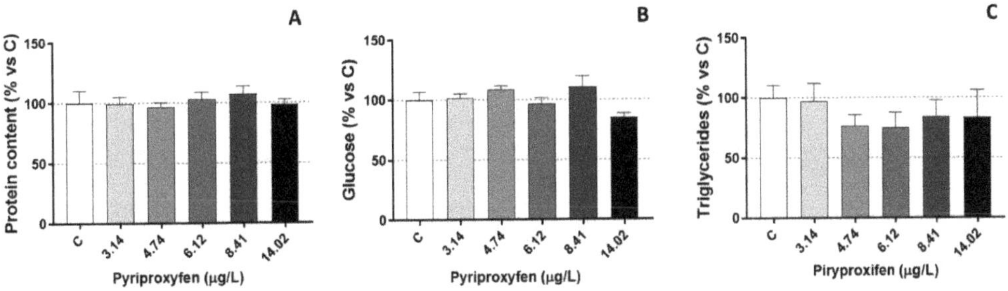

Figure 2. Protein content (**A**), glucose levels (**B**), and triglyceride levels (**C**) in *D. magna* exposed to pyriproxyfen for 21 days, compared (%) to control group C. Results are mean ± SEM.

The *fabd* gene, linked to lipid metabolism (Figure 3), showed a significant under-expression, with a reduction exceeding 40% compared to the control, in those daphnids which were exposed to all the selected pyriproxyfen concentrations. So, a drop in this gene expression would lead to an alteration in the storage of lipid reserves, which were not found in the present study.

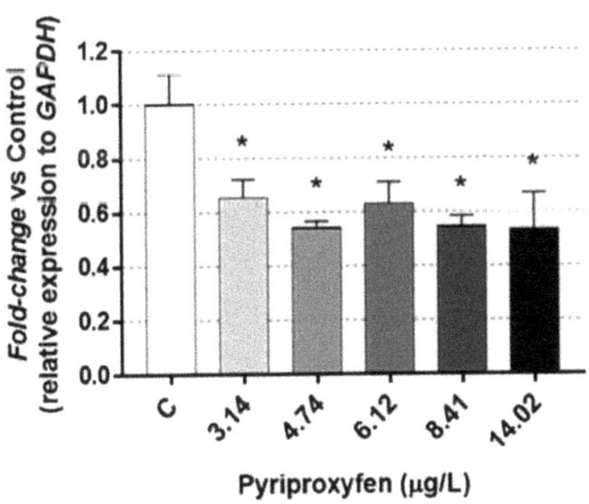

Figure 3. Results in *fabd* gene expression in *D. magna* exposed to pyriproxyfen for 21 days, compared (%) to control group C. Results are mean ± SEM (* $p < 0.05$).

Very little is known about the mechanisms by which toxicants can disrupt lipid accumulation in invertebrates. Knowledge about invertebrate lipid metabolism and their regulation needs further studies, as suggested by Fuertes et al. [43], in *D. magna*.

Some other studies have reported the overexpression of this gene in *Daphnia magna* because of exposure to substances like triphenyl phosphate [44]. In contrast, Seyoum et al. [38] investigated lipid metabolism and related genes in *D. magna* under phthalate exposure and found that the expression of *fabd* remained unchanged. These varying responses underscore the complexity of how different chemicals can influence gene expression in *D. magna*.

Moreover, it is noteworthy that pesticides can influence the structural integrity of gills, potentially reducing their efficiency as respiratory organs. In the current study, an increase in *Daphnia magna* hemoglobin (*hgb1* and *hgb2)* gene expression was observed after exposure to the highest pyriproxyfen concentrations (Figure 4). In fact, at the highest concentration tested, a significant ($p < 0.05$) overexpression, exceeding a five-fold change, was identified. This insecticide induced the overexpression of *hgb1* and *hgb2* gene products. Such inductions could function as a defense mechanism, as described by Eytcheson and LeBlanc [45]. The authors demonstrated that exposing *Daphnia magna* to a low concentration of pyriproxyfen (96 ng/L) for 48 h triggered the activation of hemoglobin synthesis. This adaptive response assisted in reducing the toxicity of sodium nitrite present in the environment, acting as a protective measure against this pollutant.

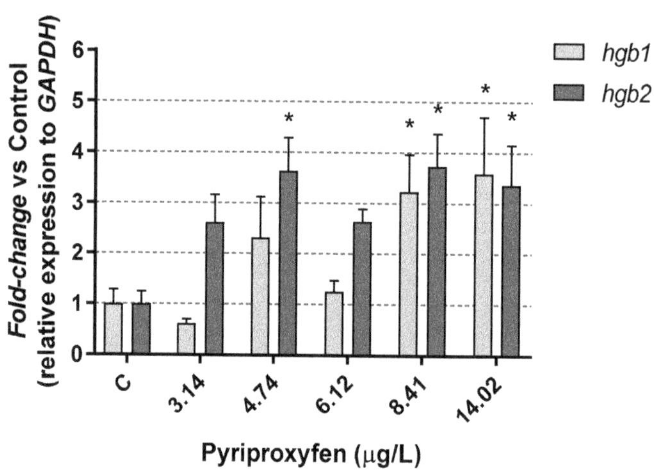

Figure 4. Hemoglobin *hgb1* and *hgb2* gene expression results in *D. magna* exposed to pyriproxyfen for 21 days, compared (%) to control group C. Results are mean ± SEM (* $p < 0.05$).

It is well-established that stress represents a mechanism of physiological compensation that an organism adopts in response to a challenging physical or chemical factor, in our case, the insecticide pyriproxyfen. Our findings are like those found by Rider et al. [28], who exposed *Daphnia magna* and *Daphnia pulex* individuals to methyl farnesoate as well as to pyriproxyfen (at concentrations ranging from 23 to 64 ng/L). They observed the stimulation of neuroendocrine cascades, leading to an increase in hemoglobin levels. This physiological response was accompanied by a visible change in the color of the individuals, transitioning from a pale hue to shades of yellowish and reddish.

In response to exposure to the insecticide, a reduction in heat-shock protein (*hsp90)* gene expression, approximately 40%, was observed in the exposed *Daphnia*, with no changes found in *hsp70* gene expression (Figure 5). Similar results were found by Silva et al. [46] in *Daphnia magna* exposed to the pesticide carbendazim. The authors suggested that the repression of heat-shock proteins might indicate that the mechanisms for cellular repairing damage could be compromised.

However, it is important to note that other studies have reported an overexpression of the *hsp90* gene in *Daphnia magna* when exposed to the flame-retardant triphenyl phosphate for 21 days [44]. Exposure to microplastics also appears to induce the overexpression of these genes, as indicated by Liu et al. [47], in *Daphnia pulex* over the exposure period.

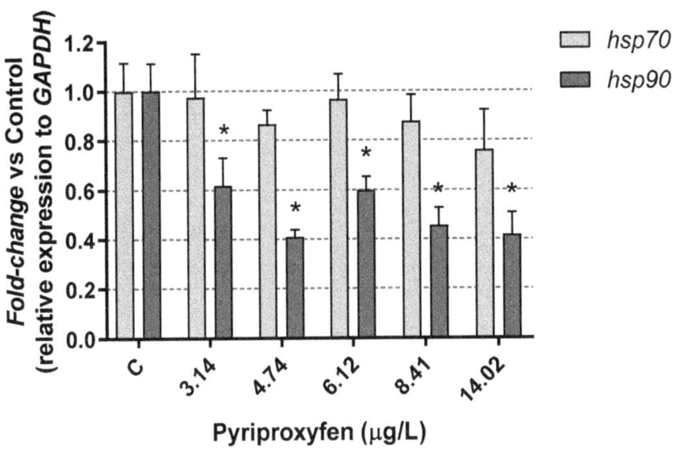

Figure 5. Heat-shock protein *hsp70* and *hsp90* gene expression results in *D. magna* exposed to pyriproxyfen for 21 days, compared (%) to control group C. Results are mean ± SEM (* $p < 0.05$).

The expression of *hsp* genes can vary not only due to the presence of xenobiotics but also due to various biological parameters, such as the presence of predators [47] or physical parameters like temperature [48]. In this context, our results are similar to those found by Kim et al. [49], who observed changes in *hsp90* expression induced by exposure to mercury. As explained by the authors, the expression patterns of these genes can fluctuate based on exposure time and other factors.

In our study, no significant ($p > 0.05$) alterations were observed in *D. magna* metallothioneins *mt-a*, *mt-b*, and *mt-c* gene expression following exposure to pyriproxyfen (Figure 6). Many studies have established a connection between the suppression of these genes with increased toxicity resulting from various types of toxic factors, including metals or pesticides. This inhibition often leads to oxidative stress and is associated with metabolic imbalances [50–52].

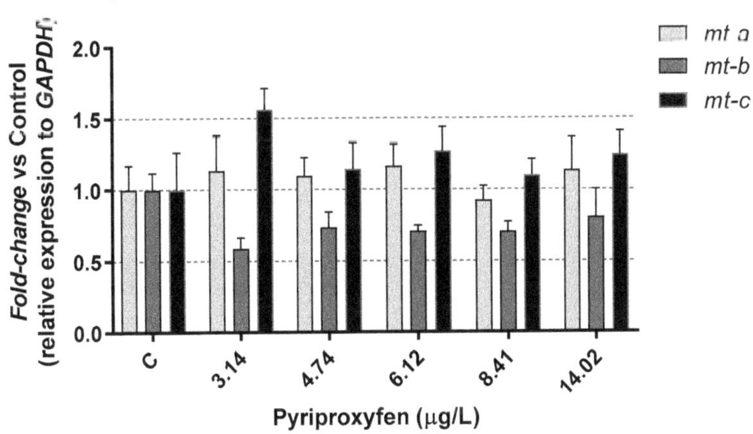

Figure 6. Metallothioneins results in *D. magna* exposed to pyriproxyfen for 21 days, compared (%) to control group C. Results are mean ± SEM.

Furthermore, products that are generally characterized as non-endocrine active compounds, such as perfluorinated alkyl substances, have been shown to inhibit the expression of these three genes in *Daphnia magna* [53]. These results highlight the diverse responses of Daphnia to different chemical compounds and their potential effects on gene expression.

In relation to oxidative stress, gluthathione-s-transferase (*gst)* gene and catalase (*cat)* gene were studied in *D. magna* after 21 days of exposure to pyriproxyfen. No significant ($p > 0.05$) alterations were observed in the expression of the *gst* gene, whereas the expression of the *cat* gene was significantly ($p < 0.05$) reduced (Figure 6). These genes are involved in antioxidant pathways and encode key antioxidant enzymes, including glutathione-S-transferase (gst) and catalase (cat), which are utilized by many aquatic organisms to mitigate the harmful effects of reactive oxygen species (ROS) [54].

It is important to note that different xenobiotics can produce varying effects on these genes in *Daphnia*. For instance, some studies have shown that phthalates inhibited the expression of both *gst* and *cat* genes in *Daphnia magna* [38,55], while the fungicide prochloraz did not significantly reduce the gene expression of *gst* and *cat* in the same species [50]. On the other hand, a decrease in *gst* and an increase in *cat* genes expression were found in *D. magna* exposed to the fungicide boscalid [56]. Kim et al. [49] observed an increase in GST activity in *Daphnia pulex* at a concentration of 2 μg/L pyriproxyfen. In our study, we did not examine GST activity, and the unaltered expression observed could suggest that the antioxidant response remained unaffected or disrupted. Further studies would be needed to provide a more comprehensive understanding of this response.

Oxidative cellular stress takes place if the concentration of free oxygen radicals (ROS) exceeds the natural neutralizing capacity of the cell [57]. Liu et al. [45] determined that the presence of microplastics in their environment seems to affect the expression of the stress defense *gst* gene in *D. magna*. The authors indicated that this effect was an age function, with both inhibition and over expression. Inhibition would be in relation to juveniles of 7 and 14 days for the gene *gst*, while overexpression would relate to early individuals (24–96 h) and the oldest individuals (21 days). In the same study, the authors observed that the *cat* gene experienced overexpression, but only after 21 days of exposure. In our results, the expression of *cat* gene was inhibited in daphnids after 21 days of exposure at all the toxicant concentrations tested, as is shown in Figure 7. This suggests a potentially disruptive effect on the response to oxidative stress.

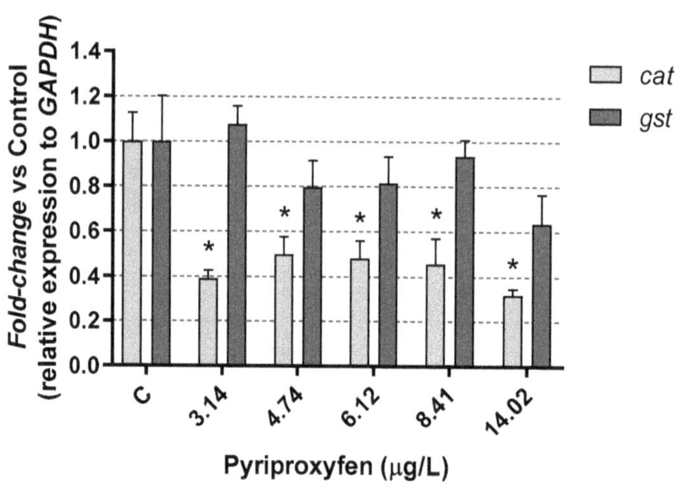

Figure 7. Oxidative stress (*cat* and *gst*) results in *D. magna* exposed to pyriproxyfen for 21 days, compared (%) to control group C. Results are mean ± SEM (* $p < 0.05$).

Vitellogenin is considered a potential early biomarker which may precede reproductive effects in Daphnia exposed to various xenobiotics. In our study, *vtg1* and *vtg2* gene expressions in *Daphnia magna* were strongly inhibited (>80%) after 21 days of pyriproxyfen exposure (Figure 8). This observation confirms the reproductive impact of pyriproxyfen on the exposed females and underscores its role as an endocrine disruptor.

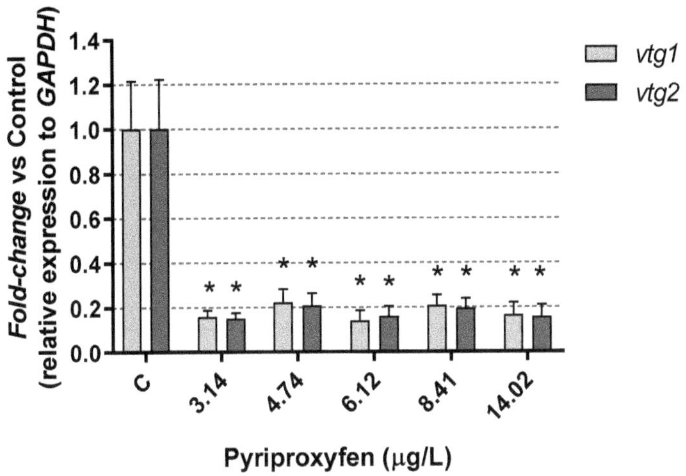

Figure 8. Vitellogenin results in *D. magna* exposed to pyriproxyfen for 21 days, compared (%) to control group C. Results are mean ± SEM (* $p < 0.05$).

These results are similar to those found by Seyoum et al. [53] in *D. magna* exposed to alkyl perfluorinated, who associated the reduction in the number of neonates per daphnia joint with the under expression of the *vtg2* gene. Vitellogenin, an essential yolk protein in Daphnia, could be stimulated by anti-ecdysteroid products [58], promoting egg development. However, the results of pyriproxyfen exposure are indicative of a strongly disruptive effect on reproduction. The lowest concentration tested, 3.14 µg/L, caused a depletion of progeny greater than 50% after 21 days, while the time to first brood was significantly increased; additionally, male progeny was observed at all exposure concentrations [35].

4. Conclusions

The exposure of *Daphnia magna* to pyriproxyfen altered its metabolism, as indicated by the differential gene expression results. However, all the animals in the study survived the exposure period at all the tested insecticide concentrations. Our findings highlight a clear endocrine-disruptive effect on *D. magna*, not only at the reproductive level, as indicated by the expression of *vtg1* and *vtg2* genes, but also with implications for the maintenance and growth of the individuals. Therefore, we could consider vitellogenin as a potential early biomarker for pyriproxyfen in *Daphnia magna*.

To obtain a better understanding of the minimal effects of pyriproxyfen on aquatic invertebrates, such as daphnids, longer exposure periods and lower concentrations should be considered. These extended assays will be valuable in identifying disruptive effects in the environment and helping to protect invertebrate populations. Invertebrate species, particularly *D. magna*, have been effectively utilized in screening and testing assays for many years. Such applications should now play a crucial role in screening and testing programs for endocrine-disrupting chemicals (EDCs).

Author Contributions: Conceptualization, B.S., E.S. and M.D.F.-R.; methodology, B.S., E.S. and M.D.F.-R.; software, B.S.; formal analysis, B.S.; investigation, B.S.; resources, E.S., M.D.F.-R. and J.T.-G.; data curation, B.S.; writing—original draft preparation, B.S. and E.S.; writing—review and editing, B.S., E.S., M.D.F.-R. and J.T.-G.; visualization, B.S.; supervision, E.S. and M.D.F.-R. All authors have read and agreed to the published version of the manuscript.

Funding: This research received no external funding.

Institutional Review Board Statement: Not applicable.

Informed Consent Statement: Not applicable.

Data Availability Statement: All data generated or analyzed in this study are included in this published article.

Conflicts of Interest: The authors declare no conflicts of interest.

References

1. Lima, E.P.; Paiva, M.H.S.; de Araújo, A.P.; da Silva, E.V.G.; da Silva, U.M.; de Oliveira, L.N.; de Melo Santos, M.A.V. Insecticide resistance in *Aedes aegypti* populations from Ceará, Brazil. *Parasit. Vectors* **2011**, *4*, 5. [CrossRef]
2. USEPA. Draft list of initial pesticide active ingredients and pesticide inerts to be considered for screening under the federal food, drug, and cosmetic act U.S. Environmental Protection Agency. In *Draft List of Chemicals for Initial Tier 1 Screening*; USEPA: Washington, DC, USA, 2007; p. 18.
3. Meola, R.; Meier, K.; Dean, S.; Bhaskaran, G. Effect of pyriproxyfen in the blood diet of cat fleas on adult survival, egg viability, and larval development. *J. Med. Entomol.* **2000**, *37*, 503–506. [CrossRef] [PubMed]
4. Wang, H.Y.; Olmstead, A.W.; Li, H.; LeBlanc, G.A. The screening of chemicals for juvenoid-related endocrine activity using the water flea *Daphnia magna*. *Aquat. Toxicol.* **2005**, *74*, 193–204. [CrossRef]
5. Ohba, S.Y.; Ohashi, K.; Pujiyati, E.; Higa, Y.; Kawada, H.; Mito, N.; Takagi, M. The effect of pyriproxyfen as a "population growth regulator" against *Aedes albopictus* under semi-field conditions. *PLoS ONE* **2013**, *8*, e67045. [CrossRef]
6. Wang, C.Y.; Teng, H.J.; Lee, S.J.; Lin, C.; Wu, J.W.; Wu, H.S. Efficacy of various larvicides against Aedes aegypti immatures in the laboratory. *Jpn. J. Infect. Dis.* **2013**, *66*, 341–344. [CrossRef]
7. Maoz, D.; Ward, T.; Samuel, M.; Müller, P.; Runge-Ranzinger, S.; Toledo, J.; Horstick, O. Community effectiveness of pyriproxyfen as a dengue vector control method: A systematic review. *PLoS Negl. Trop. Dis.* **2017**, *11*, e0005651. [CrossRef]
8. WHO. Pyriproxyfen in drinking-water: Use for vector control in drinking-water sources and containers. In *Geneva: Background Document for Development of WHO Guidelines for Drinking-Water Quality*; WHO: Geneva, Switzerland, 2008.
9. Mulla, M.S.; Darwazeh, H.A.; Kennedy, B.; Dawson, D.M. Evaluation of new insect growth regulators against mosquitoes with notes on nontarget organisms. *J. Am. Mosq. Control. Assoc.* **1986**, *2*, 314–320.
10. Schaefer, C.H.; Miura, T.; Dupras, E.F.; Mulligan, F.S.; Wilder, W.H. Efficacy, nontarget effects, and chemical persistence of S-31183, a promising mosquito (Diptera: Culicidae) control agent. *J. Econ. Entomol.* **1988**, *81*, 1648–1655. [CrossRef]
11. Schaefer, C.H.; Dupras, E.F.J.; Mulligan, F.S. Studies on the environmental persistence of S-31183 (Pyriproxyfen): Adsorption onto organic matter and potential for leaching through soil. *Ecotoxicol. Environ. Saf.* **1991**, *21*, 207–214. [CrossRef]
12. Sullivan, J.J.; Goh, K.S. Environmental fate and properties of pyriproxyfen. *Pestic. Sci.* **2008**, *33*, 339–350. [CrossRef]
13. Katagi, T.; Takahashi, N. *Hydrolysis of S-31183 in Buffered Aqueous Solutions*; Sumitomo Chemical Company: Tokyo, Japan, 1994.
14. Ccanccapa, A.; Masiá, A.; Navarro-Ortega, A.; Picó, Y.; Barceló, D. Pesticides in the Ebro River basin: Occurrence and risk assessment. *Environ. Pollut.* **2016**, *211*, 414–424. [CrossRef] [PubMed]
15. Moura, J.A.S.; Souza-Santos, L.P. Environmental risk assessment (ERA) of pyriproxyfen in non-target aquatic organisms. *Aquat. Toxicol.* **2020**, *222*, 105448. [CrossRef] [PubMed]
16. Truong, L.; Gonnerman, G.; Simonich, M.T.; Tanguay, R.L. Assessment of the developmental and neurotoxicity of the mosquito control larvicide, pyriproxyfen, using embryonic zebrafish. *Environ. Pollut.* **2016**, *218*, 1089–1093. [CrossRef]
17. Lawler, S.P. Environmental safety review of methoprene and bacterially-derived pesticides commonly used for sustained mosquito control. *Ecotoxicol. Environ. Saf.* **2017**, *139*, 335–343. [CrossRef]
18. Azevedo-Linhares, M.; Souza, A.T.C.; Lenz, C.A.; Leite, N.F.; Brito, I.A.; Folle, N.M.T.; Oliveira Ribeiro, C.A. Microcystin and pyriproxyfen are toxic to early stages of development in *Rhamdia quelen*: An experimental and modelling study. *Ecotoxicol. Environ. Saf.* **2018**, *166*, 311–319. [CrossRef]
19. Maharajan, K.; Muthulakshmi, S.; Nataraj, B.; Ramesh, M.; Kadirvelu, K. Toxicity assessment of pyriproxyfen in vertebrate model zebrafish embryos (*Danio rerio*): A multi biomarker study. *Aquat. Toxicol.* **2018**, *196*, 132–145. [CrossRef]
20. Maharajan, K.; Muthulakshmi, S.; Karthik, C.; Nataraj, B.; Nambirajan, K.; Hemalatha, D.; Ramesh, M. Pyriproxyfen induced impairment of reproductive endocrine homeostasis and gonadal histopathology in zebrafish (*Danio rerio*) by altered expression of hypothalamus-pituitary-gonadal (HPG) axis genes. *Sci. Total Environ.* **2020**, *735*, 139496. [CrossRef]
21. Gaedke, U.; Straile, D. Daphnids: Keystone species for the pelagic food web structure and energy flow. A body size-related analysis linking seasonal changes at the population and ecosystem levels. *Adv. Limnol.* **1998**, *53*, 587–610.
22. Nichols, J.; Breen, M.; Denver, R.; Distefano, J.; Edwards, J.; Hoke, R.; Volz, D.; Zhang, X. Predicting chemical impacts on vertebrate endocrine systems. *Environ. Toxicol. Chem.* **2011**, *30*, 39–51. [CrossRef]
23. Chen, W.; Fu, X.; Dong, B.; Wang, Y.; Shiah, S.; Moore, D.; Huang, W. Neonatal activation of the nuclear receptor CAR results in epigenetic memory and permanent change of drug metabolism in mouse liver. *Hepatology* **2012**, *56*, 1499–1509. [CrossRef]
24. Ginjupalli, G.K.; Baldwin, W.S. The time- and age-dependent effects of the juvenile hormone analog pesticide, pyriproxyfen on *Daphnia magna* reproduction. *Chemosphere* **2013**, *92*, 1260–1266. [CrossRef]
25. Olmstead, A.W.; LeBlanc, G.A. Insecticidal juvenile hormone analogs stimulate the production of male offspring in the crustacean *Daphnia magna*. *Environ. Health Perspect.* **2003**, *111*, 919–924. [CrossRef]

26. Sancho, E.; Villarroel, M.J.; Andreu, E.; Ferrando, M.D. Disturbances in energy metabolism of *Daphnia magna* after exposure to tebuconazole. *Chemosphere* **2009**, *74*, 1171–1178. [CrossRef]
27. De Coen, W.M.; Janssen, C.R. The missing biomarker link: Relationships between effects on the cellular energy allocation biomarker of toxicant-stressed *Daphnia magna* and corresponding population characteristics. *Environ. Toxicol. Chem.* **2003**, *22*, 1632–1641.
28. Rider, C.V.; Gorr, T.A.; Olmstead, A.W.; Wasilak, B.A.; LeBlanc, G.A. Stress signaling: Coregulation of hemoglobin and male sex determination through a terpenoid signaling pathway in a crustacean. *J. Exp. Biol.* **2005**, *208*, 15–23. [CrossRef]
29. Muyssen, B.T.; Janssen, C.R. Multigeneration zinc acclimation and tolerance in *Daphnia magna*: Implications for water-quality guidelines and ecological risk assessment. *Environ. Toxicol. Chem.* **2001**, *20*, 2053–2060.
30. OECD (Organization for Economic Co-operation and Development). *Test No. 211: Daphnia magna Reproduction Test, OECD Guidelines for Testing of Chemicals, Section 2*; OECD Publishing: Paris, France, 2012.
31. Bischoff, H.W.; Bold, H.C. Physiological studies IV. In *Some Algae from Enchaded Rock and Relate Algae Species*; Publication No. 6318; University of Texas: Austin, TX, USA, 1963.
32. Ferrando, M.D.; Sancho, E.; Andreu-Moliner, E. Effects of lindane on *Daphnia magna* during chronic exposure. *J. Environ. Sci. Health B* **1995**, *30*, 815–825. [CrossRef]
33. Sancho, E.; Villarroel, M.J.; Ferrando, M.D. Assessment of chronic effects of tebuconazole on survival, reproduction and growth of *Daphnia magna* after different exposure times. *Ecotoxicol. Environ. Saf.* **2016**, *124*, 10–17. [CrossRef]
34. Ferrando, M.D.; Fernández-Casalderrey, A.; Andreu, E. Relative sensitivity of *Daphnia magna* and *Brachionus calyciflorus* to five pesticides. *J. Environ. Sci. Health B* **1992**, *27*, 511–522. [CrossRef]
35. Salesa, B.; Torres-Gavilá, J.; Sancho, E.; Ferrando, M.D. Multigenerational effects of the insecticide Pyriproxyfen and recovery in *Daphnia manga*. *Sci. Total Environ.* **2023**, *886*, 164013. [CrossRef]
36. Dom, N.; Penninck, M.; Knapen, D.; Blust, R. Discrepancies in the acute versus chronic toxicity of compounds with a designated narcotic mechanism. *Chemosphere* **2012**, *87*, 742–749. [CrossRef]
37. Heckmann, L.H.; Connon, R.; Hutchinson, T.H.; Maund, S.J.; Sibly, R.M.; Callaghan, A. Expression of target and references genes in *Daphnia magna* exposed to ibuprofen. *BMC Genom.* **2006**, *7*, 175. [CrossRef]
38. Seyoum, A.; Pradhan, A. Effect of phthalates on development, reproduction, fat metabolism and lifespan in *Daphnia magna*. *Sci. Total Environ.* **2019**, *654*, 969–977. [CrossRef]
39. Zeis, B.; Becher, B.; Goldmann, T.; Clark, R.; Vollmer, E.; Bölke, B.; Paul, R.J. Differential haemoglobin gene expression in the crustacean *Daphnia magna* exposed to different oxygen partial pressures. *Biol. Chem.* **2003**, *384*, 1133–1145. [CrossRef] [PubMed]
40. Wu, R.S.S.; Lam, P.K.S. Glucose-6-phosphate dehydrogenase and lactate dehydrogenase in the green-lipped mussel (*Perna viridis*): Possible biomarkers for hypoxia in the marine environment. *Water Res.* **1997**, *31*, 2797–2801. [CrossRef]
41. Van Der Thillart, G. Energy metabolism of swimming trout (*Salmo gairdneri*). Oxidation rates of palmitate, glucose, lactate, alanine, leucine and glutamate. *J. Comp. Physiol. B* **1986**, *156*, 511–520. [CrossRef]
42. Villarroel, M.J.; Sancho, E.; Andreu-Moliner, E.; Ferrando, M.D. Biochemical stress response in tetradifon exposed *Daphnia magna* and its relationship to individual growth and reproduction. *Sci. Total Environ.* **2009**, *407*, 5537–5542. [CrossRef]
43. Fuertes, I.; Jordao, R.; Piña, B.; Barata, C. Time-dependent transcriptomic responses of *Daphnia magna* exposed to metabolic disruptors that enhanced storage lipid accumulation. *Environ. Pollut.* **2019**, *249*, 99–108. [CrossRef]
44. Yuan, S.; Li, H.; Dang, Y.; Liu, C. Effects of triphenyl phosphate on growth, reproduction and transcription of genes of *Daphnia magna*. *Aquat. Toxicol.* **2018**, *195*, 58–66. [CrossRef]
45. Eytcheson, S.A.; LeBlanc, G.A. Hemoglobin levels modulate nitrite toxicity to *Daphnia magna*. *Sci. Rep.* **2018**, *8*, 7172. [CrossRef]
46. Silva, R.; Silva, P.V.; Soares, A.; González-Alcaraz, M.; van Gestel, C.A.M.; Roelofs, D.; Moura, G.; Loureiro, S. *Daphnia magna* Multigeneration exposure to carbendazim: Gene transcription responses. *Toxics* **2023**, *11*, 918. [CrossRef] [PubMed]
47. Liu, Z.; Cai, M.; Yu, P.; Chen, M.; Wu, D.; Zhang, M.; Zhao, Y. Age-dependent survival, stress defense, and AMPK in *Daphnia pulex* after short-term exposure to a polystyrene nanoplastic. *Aquat. Toxicol.* **2018**, *204*, 1–8. [CrossRef]
48. Mikulski, A.; Bernatowicz, P.; Grzesiuk, M.; Kloc, M.; Pijanowska, J. Differential levels of stress proteins (HSPs) in male and female *Daphnia magna* in response to thermal stress: A consequence of sex-related behavioral differences? *J. Chem. Ecol.* **2011**, *37*, 670–676. [CrossRef]
49. Kim, H.; Kim, J.S.; Lee, Y.M. Changes in activity and transcription of antioxidant enzymes and heat shock protein 90 in the water flea, *Daphnia magna* exposed to mercury. *J. Toxicol. Environ. Health Sci.* **2017**, *9*, 300–308. [CrossRef]
50. Sato, M.; Kawakami, T.; Kondoh, M.; Takiguchi, M.; Kadota, Y.; Himeno, S.; Suzuki, S. Development of high-fat-diet-induced obesity in female metallothionein-null mice. *FASEB* **2010**, *24*, 2375–2384. [CrossRef]
51. Rodríguez-Menéndez, S.; García, M.; Fernández, B.; Álvarez, L.; Fernández-Vega-Cueto, A.; Coca-Prados, M.; González-Iglesias, H. The zinc-metallothionein redox system reduces oxidative stress in retinal pigment epithelial cells. *Nutrients* **2018**, *10*, 1874. [CrossRef]
52. Salesa, B.; Torres-Gavilá, J.; Ferrando-Rodrigo, M.D.; Sancho, E. Gene expression study alerted to possible impairment in *Daphnia magna* individuals as a consequence of exposure to sublethal concentrations of prochloraz. *Chemosphere* **2022**, *308*, 136040. [CrossRef]
53. Seyoum, A.; Pradhan, A.; Jass, J.; Olsson, P.E. Perfluorinated alkyl substances impede growth, reproduction, lipid metabolism and lifespan in *Daphnia magna*. *Sci. Total Environ.* **2020**, *737*, 139682. [CrossRef]

54. Yu, P.; Liu, Z.; Wu, D.; Chen, M.; Lv, W.; Zhao, Y. Accumulation of polystyrene microplastics in juvenile *Eriocheir sinensis* and oxidative stress effects in the liver. *Aquat. Toxicol.* **2018**, *200*, 28–36. [CrossRef]
55. Wang, Y.; Wang, T.; Ban, Y.; Shen, C.; Shen, Q.; Chai, X.; Wei, J. Di-(2-ethylhexyl) phthalate exposure modulates antioxidant enzyme activity and gene expression in juvenile and adult *Daphnia magna*. *Arch. Environ. Contam. Toxicol.* **2018**, *75*, 145–156. [CrossRef]
56. Aksakal, F.I. Evaluation of boscalid toxicity on *Daphnia magna* by using antioxidant enzyme activities, the expression of genes related to antioxidant and detoxification systems, and life-history parameters. *Comp. Biochem. Physiol. Part C* **2020**, *237*, 108830. [CrossRef]
57. Flora, S.J.S. Structural, chemical and biological aspects of antioxidants for strategies against metal and metalloid exposure. *Oxid. Med. Cell Longev.* **2009**, *2*, 191–206. [CrossRef]
58. Hannas, B.R.; Wang, Y.H.; Thomson, S.; Kwon, G.; Li, H.; Leblanc, G.A. Regulation and dysregulation of vitellogenin mRNA accumulation in daphnids (*Daphnia magna*). *Aquat. Toxicol.* **2011**, *101*, 351–357. [CrossRef]

Journal of *Xenobiotics*

MDPI

Article

Exploring the Effects of Graphene-Based Nanoparticles on Early Salmonids Cardiorespiratory Responses, Swimming and Nesting Behavior

Tomas Makaras [1,*], Magdalena Jakubowska-Lehrmann [2], Živilė Jurgelėnė [1] and Sergej Šemčuk [3]

1 Nature Research Centre, Akademijos St. 2, 08412 Vilnius, Lithuania; zivile.jurgelene@gamtc.lt
2 National Marine Fisheries Research Institute, Kołłątaja St. 1, 81-332 Gdynia, Poland; mjakubowska@mir.gdynia.pl
3 Center for Physical Sciences and Technology, Saulėtekio Av. 3, 02300 Vilnius, Lithuania; sergej.semcuk@ftmc.lt
* Correspondence: tomas.makaras@gamtc.lt

Abstract: Graphene-based nanomaterials are exceptionally attractive for a wide range of applications, raising the likelihood of the release of graphene-containing nanoparticles into aquatic environments. The growing use of these carbon nanomaterials in different industries highlights the crucial need to investigate their environmental impact and evaluate potential risks to living organisms. The current investigation evaluated the nanotoxicity of graphene (nanoflakes) and graphene oxide (GO) nanoparticles on the cardiorespiratory responses (heart rate, gill ventilation frequency), as well as the swimming and nesting behavioral parameters of early stage larvae and juvenile salmonids. Both short-term (96 h) and long-term (23 days) exposure experiments were conducted using two common species: brown trout (*Salmo trutta*) and rainbow trout (*Oncorhynchus mykiss*). The findings demonstrated notable alterations in fish nesting behavior, swimming performance, and cardiorespiratory functions, indicating the potential toxicity of nanoparticles. This impact was observed at both physiological and whole-organismal levels in salmonids at early stages. Future investigations should explore different types of nanocarbons and their potential enduring effects on fish population structure, considering not only individual survival but also broader aspects of development, including feeding, reproductive, and other social dynamics.

Keywords: gill ventilation; heart rate; larval stage; juveniles; nests; swimming activity

1. Introduction

Carbon nanomaterials, including graphene, which have been discovered in the last several years, exhibit remarkable properties with significant potential across a spectrum of applications [1,2]. These applications encompass energy storage, electronics, nanocomposites, medical applications, extreme condition systems, desalination, antifouling agents, and the purification of pollutants from seawater. The demand for such nanocarbons is expected to exceed 1 kiloton annually by the early 2025s, with additional growth foreseen as the nanotechnology sector progresses [3]. The increasing utilization of nanocarbons, combined with the lack of regulatory constraints and efficient waste management, presents a growing threat to both the environment and human safety [4]. Therefore, it is essential to meticulously define and establish scientifically valid thresholds for the presence of nanomaterials in aquatic environments, ensuring the sustainability and welfare of aquatic species by considering factors such as size, composition, and concentration.

The existing or anticipated environmental concentrations of carbon nanomaterials are unclear because of inadequate measurement methods. Due to limitations in analytical methodologies, the fate of these nanomaterials in the environment remains unknown.

While the technological processes for synthesizing various graphene-based nanomaterials (GBNs) may be similar, slight variations in the properties of the final products can lead to diverse induced toxicity effects. Thus, when identifying adverse effects on aquatic biota from a particular material, it is essential to comprehensively understand not only the carbon nanomaterials themselves but also to attribute these effects to their physicochemical characteristics. The toxicity of GBNs might be influenced by specific characteristics such as linear size, particle form, quantity of layers, surface characteristics, functionalization, impurities, and their susceptibility for aggregation, clustering, and sedimentation [5,6]. The dynamic interaction of GBNs with living organisms, spanning a size range from a few to 100 nm, stands as a critical factor influencing their compatibility within biological systems [7]. Whether taken in by organisms or adhering through biological barriers such as skin, mucosal membranes, and the blood–brain barrier, GBNs harbor the potential to instigate mechanical damage to cells, thereby disrupting diverse cellular functions [8]. These disturbances extend to the induction of oxidative stress, the alteration of cell viability, and the modulation of immunological, physiological, and behavioral responses, collectively impacting the overall health of the organism [9–11].

Recent studies have revealed that graphene oxide (GO) can induce genotoxic, immunotoxic, and cytotoxic changes in living organisms, including enhanced apoptosis, oxidative stress, cardiac and metabolic impairments, and disruptions in fish embryo development, all of which have implications for their survival and population structure [12–15]. The toxic mechanisms of GO-induced toxicity in fish have also been extensively examined at the cellular and molecular levels [16,17]. However, most of the studies demonstrating the adverse effects in fish focus on the crucial role that fish play in the transport of nanomaterials within and beyond water body boundaries. The success of population recruitment is heavily dependent on the well-being of these early life stages. Many of these studies were primarily conducted in vivo using zebrafish (*Danio rerio*) [12,13,18,19], or a Japanese medaka (*Oryzias latipes*) which are known as a model species characterized by a relatively brief embryonic and larval development [20,21]. Nevertheless, the impact of carbon nanomaterials on the initial phases of different fish species, such as salmonids, which undergo an extended embryonic development, remains uncertain. While salmonids are an economically valuable fish species that naturally inhabit aquatic environments and may be directly exposed to various nanocarbons, there is a scarcity of toxicological data pertaining to these species. Information on induced effects at different stages of development and across various biological levels is lacking. Therefore, further exploration is warranted across a spectrum of scales, from cellular to whole-organism levels.

To enhance understanding and supplement existing data on the effects of GBNs on freshwater salmonids, this study examined the short-term (4 days) and long-term (23 days) impacts of graphene (nanoflakes) and GO on brown trout (*Salmo trutta*) and rainbow trout (*Oncorhynchus mykiss*) at various developmental stages (larvae and juveniles). Utilizing the popular form of graphene (nanoflakes) and the more economical and extensively studied GO as its precursor, we assessed physiological parameters such as heart rate (HR) and gill ventilation frequency (GVF), along with nesting rate and swimming activity. These analyses aimed to elucidate the influence of these carbon nanomaterials at both physiological and whole-organismal levels, providing insights into early stage fish behavioral traits that could impact their survival.

2. Materials and Methods

2.1. Study Animal

Brown trout (*S. trutta*) and rainbow trout (*O. mykiss*) eyed-stage embryos were obtained from fish hatcheries in Trakų Vokė, Lithuania, and Dąbie, Poland. Rainbow trout embryos followed the incubation protocol as described in [22]. In 1.7 L glass containers, 300 rainbow trout eggs were placed, along with 1.5 L of freshwater (filtered through GF/F filters; Whatman 47 mm, 0.7 μm) and 200 g of gravel sediment (3 mm in diameter). These containers were incorporated into a closed-loop recirculation system equipped with a cooling unit (Titan 2000, Aqua Medic, Bissendorf, Germany) to maintain a steady water temperature of 9 ± 0.5 °C.

Brown trout eggs were put in incubators (430 × 510 × 150 mm) made of porous stainless steel (2 mm in diameter), ensuring a steady water flow at a temperature of 8 ± 0.5 °C. Brown trout juveniles, originating from a stock population within the same spawning, were accommodated in Recirculating Aquaculture Systems (RAS) in 1 m^{-3} capacity tanks. The flow-through system, with a water density below 25 kg m^{-3}, maintained a flow rate of 10–15 mm s^{-1}, and the environment was controlled at a pH of 8 and a temperature of 12 °C, following natural dark–light cycles.

Experimental trials were conducted on larval stages (1-day post-hatching) for both *S. trutta* and *O. mykiss*, as well as on *S. trutta* juveniles (~8 months old, 0+) with a weight (mean ± SD) of 5.91 ± 1.42 g and a length of 6.51 ± 0.72 cm (n = 80).

All biological testing followed the guidelines specified in [23] for the ethical treatment of animals used in scientific research. The protocol received approval from the Animal Ethics Committees of both the Lithuanian State Food Veterinary Service (license no. G2-168; valid from 8 February 2021 to 7 February 2025).

2.2. Synthesis and Characterization of Graphene Oxide

The fresh portion of GO was synthesized based on methodology described in our previous study [24]. The synthesized GO was subjected to thorough characterization via Fourier-transform infrared spectroscopy (FTIR), X-ray powder diffraction (XRD), Raman spectroscopy, thermal gravimetric analysis (TGA), X-ray fluorescence (XRF), and scanning electron microscopy (SEM) analysis.

The FTIR spectra of prepared GO were recorded with ALPHA spectrometer (Bruker Inc., Ettlingen, Germany) from 400 to 3800 cm^{-1} with resolution of 4 cm^{-1} to affirm the formation of functional groups: carboxyl, epoxy, and hydroxyl. The recorded spectra exhibited peaks at 1053, 1080, 1230 1420, 1621, 1736 cm^{-1} and one wide 3000–3300 cm^{-1}, corresponding to C–O, C–O–C, C–OH, C–C, C=O, and O–H vibrations, respectively. The Raman spectra was recorded using RAM II system (Bruker Inc., Ettlingen, Germany) with a 1064 nm laser. The D ang G bands were detected at ranges 1300–1350 cm^{-1} and 1595–1610 cm^{-1}, respectively. The calculated intensity ratio of those bands was 0.76, indicating a high degree of oxidation of synthesized GO [25]. The XRD pattern was captured utilizing a RIGAKU X-ray diffractometer from Japan and provided for analysis with the PDXL software package (version 2.8.4.0). The successful synthesis of GO was validated by a significant peak at 2θ = 10.44°. Moreover, calculations of crystallite size of GO applying the Halder–Wagner method [26] showed a result of 6.4 ± 0.4 nm. The purity of produced GO was examined with TGA (STA Pt 1600 with mass spectrometer Thermostar GDS 320, Linseis, Selb, Germany) and XRF (Fluorescent X-ray spectrometer, Panalytical, Almelo, The Netherlands) techniques. Both detected the sulphur as a main impurity (potentially originating from the H_2SO_4 used during the synthesis) with 0.119% of tested samples and a purity of GO was 99.621 ± 0.011%. The size distribution was examined with SEM Helios NanoLab 650 (FEI, Eindhoven, The Netherlands). Analysis of images showed that the size of synthesized GO sheets is, on average, 10 μm, with nm scale thickness (Figure 1A,B).

The commercially available graphene (nanoflakes) were purchased from Graphene Supermarket Inc. (Ronkonkoma, New York, NY, USA). Because of natural aggregation, the thickness might range from one to a few graphitic sheets. The linear size of these sheets fluctuates between 8 and 20 nm, with a grain size distribution approximately ~12 nm (Figure 1C,D).

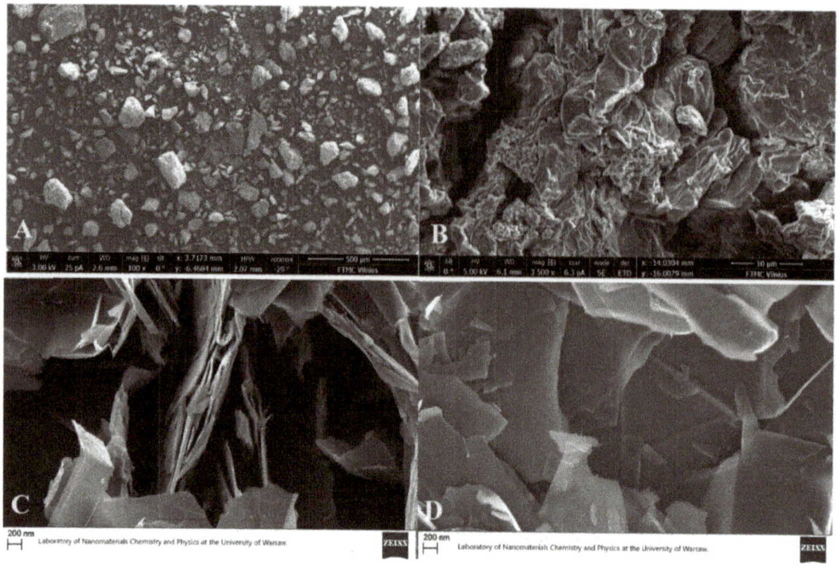

Figure 1. SEM images depicting the synthesized graphene oxide (GO) (**A**,**B**), and graphene (nanoflakes) (**C**,**D**). SEM images adapted from our previous studies [24,27].

2.3. Experimental Design

To analyze the impact of graphene (nanoflakes) and GO on brown trout (*S. trutta*) and rainbow trout (*O. mykiss*), the research was divided into short-term (up to 96 h) and long-term (up to 23 days) experiments, considering the stage of fish development and the tested parameters. A schematic representation of the experimental settings is provided in Figure 2.

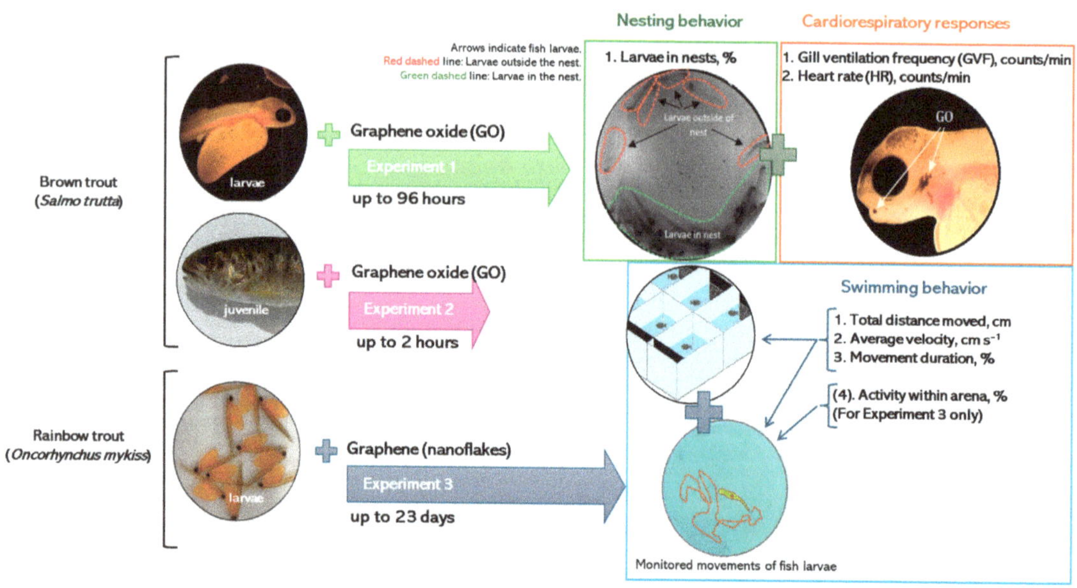

Figure 2. A flowchart of experimental settings.

2.3.1. Experiment 1: Short-Term Exposure of *S. trutta* Larvae to GO

Short-term behavioral tests were conducted on *S. trutta* larvae and juveniles, exposing them to various concentrations of GO (0, 1.0, 10.0, 20.0, and 40.0 mg L^{-1}). Fish larvae were tested in groups of 10 per replicate (40 per treatment) in 0.6 L glass beakers filled with 0.5 L of water with dispersed GO achieved through stirring or ultrasonication with EMAG EM40HC (at 38 kHz) for up to 2 min period. The control group received GO-free water from RAS.

Experiments lasted 96 h in a climate chamber (PGC-660, Bronson, Zaltbommel, The Netherlands) at a constant water temperature of 8 ± 0.5 °C and under dark lighting conditions. Continuous aeration was maintained to maximize oxygen saturation and minimize GO particle aggregation intensity.

The quantity (expressed as a percentage, %) of larvae within nests (identified when one larva closely interacts with another) was observed and computed after 24, 48, 72, and 96 h of exposure. Heart rate (HR) and gill ventilation frequency (GVF) of 5 larvae per treatment were calculated after 96 h. Cardiorespiratory monitoring was carried out after 96 h of exposure, examining each larva for 15 s in the dark to avoid light interference. No deceased larvae were observed throughout the 96 h exposure period.

Experiments with larvae were conducted in compliance with ISO standards [28] and the OECD recommendations for acute toxicity tests [29].

2.3.2. Experiment 2: Short-Term Exposure of *S. trutta* Juveniles to GO

For the analysis of swimming activity, individual *S. trutta* juveniles were placed in open-field glass tanks (350 × 200 × 200 mm), with 16 individuals per treatment. Each tank was filled with 3 L of water (13 ± 1 °C) sourced from RAS. Fish acclimatization lasted 30 min prior to testing. In exposure trials, the GO test concentration, similar to that used for exposed fish larvae, was introduced and thoroughly mixed to prevent particle aggregation.

The trials were carried out under static water conditions, signifying that the water remained unchanged until the experiment's conclusion. The water was consistently aerated. The temperatures of the water and dissolved oxygen (DO) saturation (99.7%) in the experimental tanks holding the test fish were steady throughout the study.

Swimming activity data were analyzed using Ethovision XT 16 (Noldus Inc., Wageningen, The Netherlands) video tracking software. Common endpoints such as total distance moved (cm), average velocity (cm s^{-1}), and cumulative movement duration (%) were selected for analysis. Video data were gathered at intervals of 10 min throughout the 2 h experimental periods. Fish swimming behavior was evaluated by consolidating group-mean data after a 2 h period.

Experiments 1 and 2 were carried out at the Nature Research Centre (Lithuania).

2.3.3. Experiment 3: Long-Term Exposure of *O. mykiss* Larvae to Graphene (Nanoflakes)

This study examined the prolonged impact (23 days) of graphene (nanoflakes) on the swimming behavior of *O. mykiss* larvae (220 D°) in accordance with a previous study [29]. Each individual larva was placed in transparent, cylinder-shaped plastic jar (70 × 100 mm) filled with 0.3 L of a graphene (nanoflakes) solution at a concentration of 4 mg L^{-1}, consistent with the exposure groups used for stocking (5 larvae each replication, total of 15 per treatment). Additionally, larval groups (10 individuals per replicate, 90 in total per treatment) were exposed, and their overall group activity duration, measured as pixel changes across the observational arena (activity within arena, %), was assessed using the same video tracking software. All larvae were simultaneously monitored in their respective jars.

The choice of the 4 mg L^{-1} concentration was based on the outcomes observed in a short-term experiment and the observed effects at the lowest concentration of graphene oxide (GO). It also aligns with the lower range of carbon nanomaterial concentrations commonly applied in experimental studies on fish's early life stages [12,15,30,31].

The jars were moved to a flow-through system located in an open-field glass tank to ensure a stable water temperature (9 ± 1 °C) and monitor the movement of the larvae. A 30 min acclimation period preceded the recording of swimming activity. Trials were carried out under static water conditions and natural lighting, with dissolved oxygen (DO) levels consistently maintained above 9.6 mg L^{-1}. Data were gathered at 10 min intervals during a 5 min recording period, and group-mean data after 1 h were employed for additional analysis, as outlined in Section 2.4.

Experiment 3 was carried out at the National Marine Fisheries Research Institute (Poland).

2.4. Statistical Analysis

Statistical data evaluation included checking for normality using the Shapiro–Wilk test and assessing homoscedasticity through the Levene test. Visual inspection of residuals was conducted, and when necessary, log- or square-root transformations were applied, followed by re-analysis to meet parametric assumptions. If the requirements were not met post-transformation, non-parametric tests were employed.

Nesting and short-term swimming behavioral data over time exposure underwent Repeated Measures ANOVA, with Mauchly's sphericity test ensuring model fitness. As no significant RM ANOVA model, including the time variable, was identified, the swimming activity data over time was consolidated and subjected to one-way ANOVA. Fish larval swimming data over 16-, 20-, and 23-day periods were analyzed using linear mixed-effects models (LMEMs), incorporating replicate and larvae ID (for individual behavior) as random factors and time/treatment as fixed effects. T-independent tests were used to determine significant differences between treatments at specific timeframes. For cardiorespiratory data, the non-parametric Kruskal–Wallis test was utilized to identify significant differences among independent groups. Each ANOVA was followed by a post-hoc Tukey's HSD test, with significance set at $p < 0.05$. Hierarchical cluster analysis (HCA) identified treatments with comparable effects on fish, utilizing normalized data, Euclidean distance, and average linkage.

Statistical analyses and visualizations utilized STATISTICA (v10.0, StatSoft Innovations; Tulsa, OK, USA) and PRISM (v8.0.0, GraphPad Technologies; San Diego, CA, USA) software.

3. Results

3.1. The Short-Term Effects of GO on S. trutta Larvae

3.1.1. Cardiorespiratory Functions

The examination of cardiorespiratory endpoints of brown trout (*S. trutta*) larvae after 96 h revealed that only GVF showed any slight changes, with a 9.3 ± 2.1% decrease at the highest tested concentration (Kruskal–Wallis H = 27.3, $p = 0.02$; Figure 3A). In contrast, heart rate (HR) (Figure 3B) proved to be a less sensitive endpoint, displaying insignificant changes ($p > 0.05$) compared to the control level, ranging from 84 to 104 counts/min, similar to those exposed to GO concentrations at 20.0 and 40.0 mg L^{-1}.

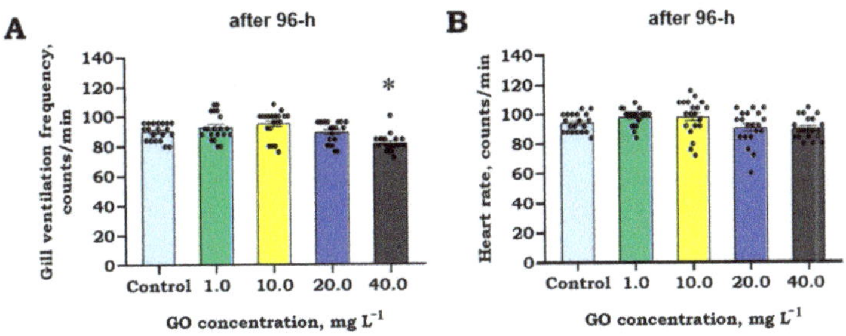

Figure 3. Gill ventilation frequency (GVF) (count/min), (**A**) and heart rate (HR) (count/min), (**B**) (mean ± SEM, raw data, n = 20) during 96 h exposure trials exposure to different concentrations of graphene oxide (GO). Asterisks (*) denote significant differences between treatments compared to the control level at significance levels of $p < 0.05$ (*).

3.1.2. Nesting Behavior

The nesting behavior of *S. trutta* larvae (Figure 4) was significantly impacted by GO over a 96 h period, revealing a noteworthy interaction between time and treatment (Repeated Measures ANOVA, $F_{12,45} = 2.2$, $p = 0.031$; Figure 4A). Following a 24 h exposure to GO concentrations of 10 (Tukey's HSD test, $p = 0.025$), 20 ($p = 0.002$), and 40 ($p < 0.001$) mg L^{-1}, a significant decrease (% change ± SEM), ranging from 10.8 ± 1.9 to 23.1 ± 4.3% of larvae in nests, was observed compared to the control level. Interestingly, as the length of GO exposure grew, the number of larvae in nests also grew, indicating a possible adaptive response to extended GO exposure.

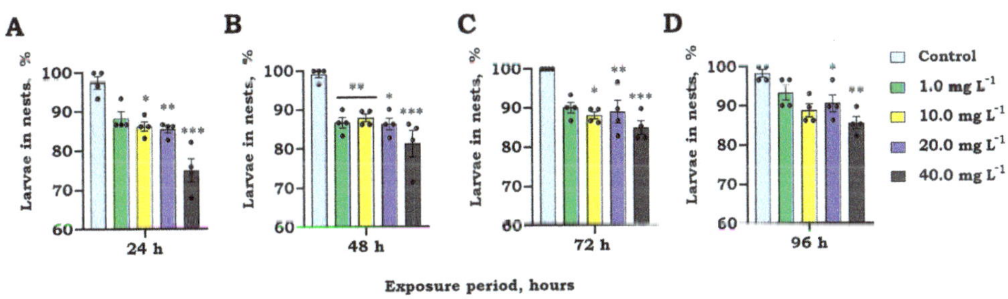

Figure 4. Number of brown trout (*S. trutta*) larval individuals (1 day after hatching) in nests (%) (**A**) (mean ± SEM, raw data, n = 4 repl.) after 24 h (**A**), 48 h (**B**), 72 h (**C**), and 96 h (**D**) exposure to different concentrations of graphene oxide (GO). Asterisks (*) denote significant differences between treatments compared to the control observational timeframe at significance levels of $p < 0.05$ (*), <0.01 (**), and <0.001 (***).

3.2. The Short-Term Effects of GO on S. trutta Juveniles

Swimming Activity

The obtained behavioral data in brown trout (*S. trutta*) juveniles revealed significant alterations in swimming activity endpoints, including total distance moved ($F_{4,75} = 8.3$, $p < 0.001$), average velocity ($F_{4,75} = 9.2$, $p < 0.001$), and movement duration ($F_{4,75} = 7.9$, $p < 0.001$), following a 2 h exposure to various concentrations of GO as illustrated in Figure 5. Analyzing these endpoints, the impact of GO on fish juveniles was characterized by a considerable increase from the control level, ranging from 53.5 ± 12 to 72.2 ± 15% ($p < 0.01$), across almost all tested concentrations (10.0, 20.0, and 40.0 mg L^{-1}) of GO.

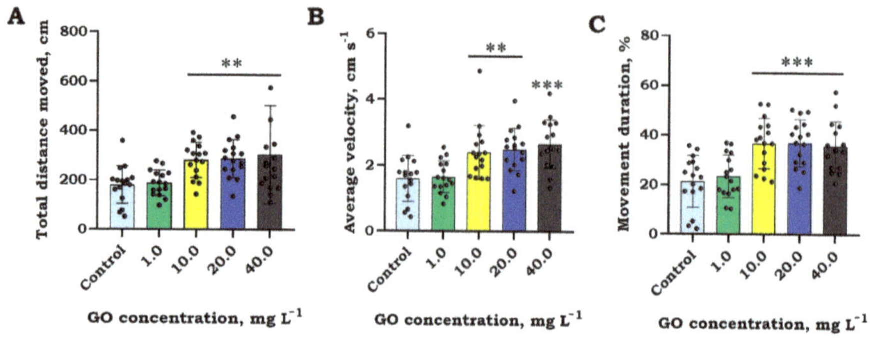

Figure 5. Total distance moved (cm) (**A**), average velocity (cm/s) (**B**), and (cumulative) movement duration (%) (**C**) (mean ± SEM, raw data, n = 16) for brown trout (*S. trutta*) juveniles after 2 h of exposure to different concentrations of graphene oxide (GO). Asterisks denote significant differences between treatments compared to the control observational timeframe at significance levels of $p < 0.01$ (**), and < 0.001 (***).

3.3. The Hierarchical Cluster Analysis

The hierarchical cluster analysis (HCA) revealed distinct groupings among the studied fish (brown trout) (Figure 6). The first cluster included *S. trutta* larvae and juveniles exposed to 1.0 mg L^{-1} GO, along with the control group, indicating similar behavioral responses. In the second major cluster, fish treated with 10.0 and 20.0 mg L^{-1} GO formed one subcluster, while those exposed to 40.0 mg L^{-1} GO comprised another. This detailed categorization highlights nuanced variations in fish responses to different GO concentrations, emphasizing concentration-dependent effects and the significance of dosage levels in shaping observed behavioral patterns.

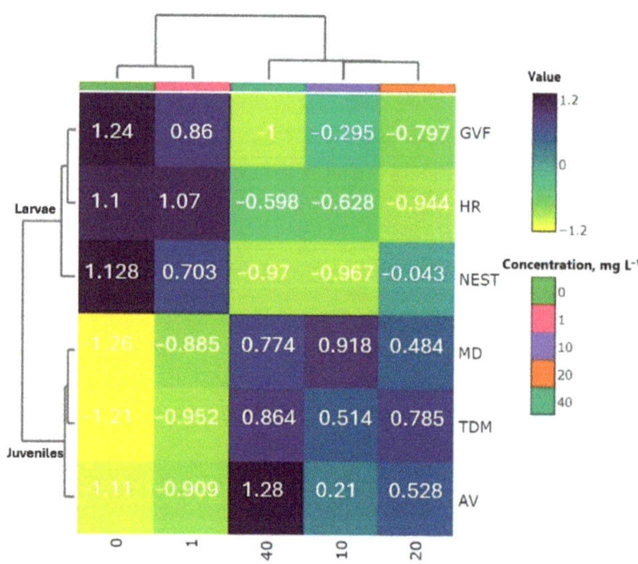

Figure 6. Hierarchical cluster analysis of brown trout (*S. trutta*) larvae and early juveniles exposed to different concentrations of GO based on the Euclidean distance coefficient and the average linkage method. Abbreviations: AV—average velocity (cm s^{-1}), TDM—total distance moved (cm), MD—movement (cumulative) duration (%), HR—heart rate (counts/min), GVF—gill ventilation frequency (counts/min), and NEST—larvae in nests (%).

3.4. The Effects of Graphene (Nanoflakes) on *O. mykiss* Larvae

Swimming Activity

Prolonged exposure of rainbow trout (*O. mykiss*) larvae to graphene (nanoflakes) at a concentration of 4.0 mg L^{-1} for 23 days did not reveal any significant interaction between time and treatment, as illustrated in Figure 7. A notable variability in individual responses was observed with an increasing duration of exposure. Nevertheless, when assessing variations in responses during specific observational timeframes, t-independent test analysis revealed significant differences in larval swimming activity, as indicated in Figure 8.

The presence of graphene nanoflakes resulted in noticeable behavioral changes on the 20th day of the exposure period, characterized by an increase in total distance moved (t-test = −3.27, p = 0.002; Figure 8A), average velocity (t-test = −3.38; p = 0.002; Figure 8B), movement (cumulative) duration (t-test = −2.44; p = 0.021; Figure 8C), as well as changes in activity within the arena following larvae exposure in a group, where activity was determined to be significantly increased (t-test= −2.19; p = 0.044; Figure 8D). Furthermore, it should be noted that during long-term exposure to graphene, larval activity significantly increased over development, contributing to the observed high variation in responses.

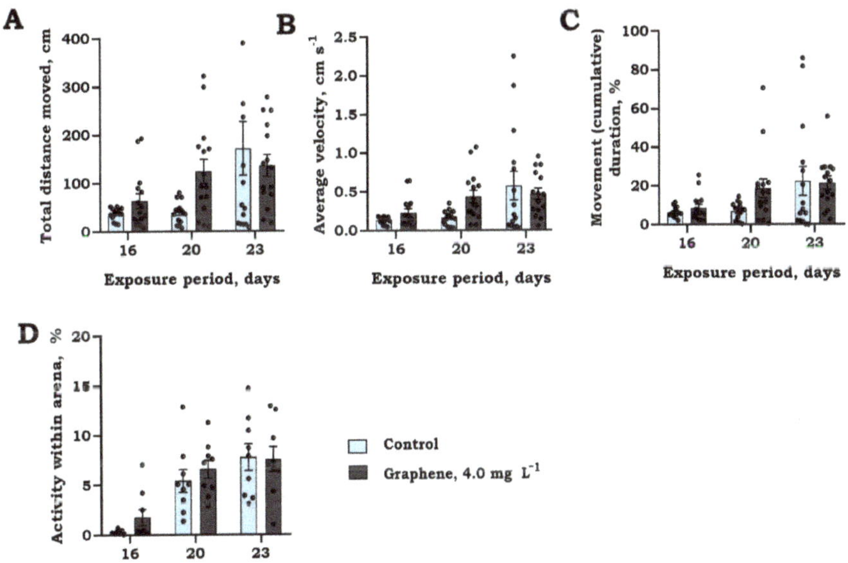

Figure 7. Total distance moved (cm) (**A**), average velocity (cm s^{-1}) (**B**), movement (cumulative) duration (%) (**C**) (mean ± SEM, raw data, n = 15), and activity within arena (%) (**D**) (mean ± SEM, n = 9 repl.) of rainbow trout (*O. mykiss*) larvae after selected time intervals (16, 20, and 23 days) of exposure to graphene (nanoflakes) at a concentration of 4.0 mg L^{-1}.

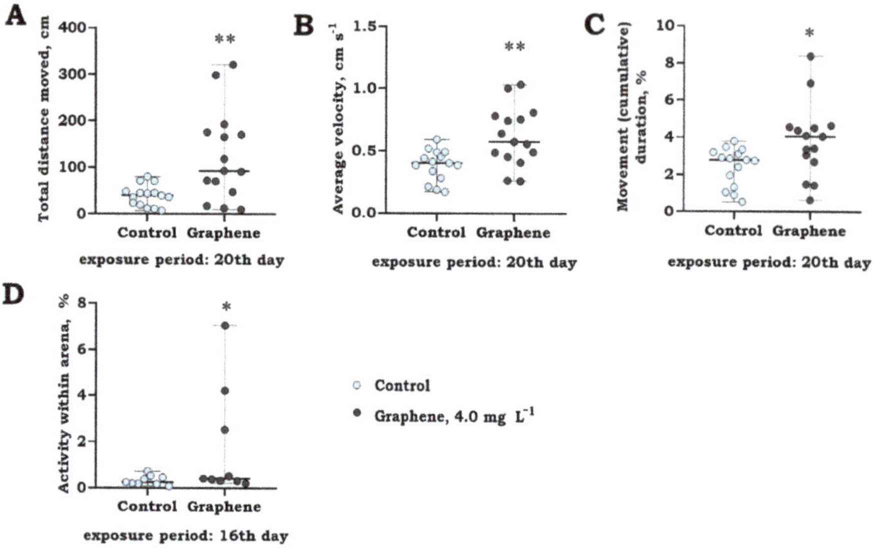

Figure 8. Total distance moved (cm) (**A**), average velocity (cm s^{-1}) (**B**), movement (cumulative) duration (%) (**C**) (median with range, raw data, $n = 15$), and activity within arena (%) (**D**) (median with range, raw data, $n = 9$ repl.) of rainbow trout (*O. mykiss*) larvae following 16, 20, and 23 days of exposure to graphene (nanoflakes) at a concentration of 4.0 mg L^{-1}. Asterisks denote significant differences between treatments compared to the control level at significance levels of $p < 0.05$ (*) and <0.01 (**).

4. Discussion

This study aimed to investigate the impact of short-term (up to 96 h) and long-term (up to 23 days) exposure to graphene, in the form of nanoflakes, and its oxidized variant, graphene oxide (GO), on the early developmental stages and juveniles of two common salmonid species: brown trout (*S. trutta*) and laboratory-reared rainbow trout (*O. mykiss*). This research aimed to assess the effects of these carbon nanomaterials at both physiological and whole-organismal levels, providing a deeper understanding of the early life behavioral traits of freshwater fish that could impact their survival rates.

In the present study, exposure of *S. trutta* larvae to GO resulted in a reduction in nesting activity by 11–23%. Ordinarily, after hatching, larvae exhibit a natural behavioral trait of forming nests, driven by an instinct to grow in gravel spawning nests for protection. When fish experience stress, their natural response is to avoid and escape, leading to a lack of nest-building and, consequently, a decreased likelihood of survival. In addition to avoidance, alternative mechanisms may be implicated, such as the secretion of epidermal mucus as a defense against adverse environmental conditions and the uptake of nanoparticles [32]. In our long-term exposure study, we observed a significant increase (up to 3.3-fold) in the swimming activity endpoints of *O. mykiss* larvae when exposed to graphene nanoflakes. As swimming behaviors are closely linked to their cardiorespiratory systems, playing a crucial role in efficient oxygen uptake [33], it was noted that cardiorespiratory functions were also impacted by GO in *S. trutta* larvae, particularly at the highest tested concentration.

Despite the significant interest among scientists in carbon nanomaterials, toxicity tests have primarily focused on a variety of model organisms, with particular attention given to zebrafish (*Danio rerio*) during their early life stages. These studies aimed to explore the negative effects of graphene-based nanomaterials on physiology, histopathology, and behavioral patterns of the whole organism [19,34,35]. The impact of graphene derivatives on the swimming performance of both fish larvae and adult fish has been demonstrated to exhibit diverse effects, including increased or decreased activity, abnormal swimming patterns, and, in some cases, no discernible effects. Due to the adverse impact of graphene derivatives on fish behavior, several studies have proposed a potential association between these carbon nanomaterials and developmental neurotoxicity as well as immunotoxicity during the early developmental stages [9,14,30,36]. Moreover, changes in fish swimming activity have been associated with variations in acetylcholine esterase and cortisol levels following treatment with the oxidized form of graphene. This impact was determined in zebrafish larvae exposed to relatively low concentrations (0.1 and 1.0 mg L^{-1}), similar to our lowest tested concentration [34]. Furthermore, the accumulation of graphene nanoparticles in the area of the head has been identified as a factor causing cellular damage in the brain, leading to degeneration and the production of autophagosomes, ultimately influencing behavior [30]. The available evidence suggests that the effects seen in fish are largely contingent on the duration of exposure, with a direct correlation to the dose and lateral size of these carbon compounds [15]. Despite various mechanisms proposed to explain changes in the behavior of fish larvae linked to the mode of action of graphene oxide, research on these nanoparticles neurotoxic effects is being extensively investigated.

Numerous investigations have primarily concentrated on evaluating developmental neurodegeneration and immune system damage caused by GO, with a primary focus on zebrafish embryos. However, it is crucial to consider and investigate the impact of GO on later developmental stages of fish. Early life fish juveniles are known to be sensitive, particularly salmonids, similar to larvae, to adverse environmental changes, including water pollution with nanocarbons. In addition to studying fish larvae, we explored the effects of GO on *S. trutta* juveniles at the age of 0+. Our study observed that *S. trutta* juveniles exposed to GO particles exhibited a substantial increase in swimming activity by 54–72% within first two hours, resembling the effects observed in *O. mykiss* larvae during long-term exposure. The adverse effects of GO on fish can be attributed to the mechanical harm inflicted on specific tissues and organs tailored to fish.

In the early stages of development, fish gills, being delicate organs and the primary point of entry for noxious substances, are susceptible to damage even in minimal concentrations. Graphene-based nanomaterials (GBNs) can cause irritation and damage to the gills by interacting with their branchial membrane or mucoproteins. This interaction may lead to acute gill irritation and a prompt avoidance response in fish, especially during the initial hours of exposure [37]. Consistent with the findings of a previous study [13], necrotic cell areas were identified in the gill tissues of zebrafish exposed to GO concentrations (10–20 mg L^{-1}, respectively), mirroring the concentrations investigated in our research. On the other hand, some authors have suggested that hyperactivity might be related to the particles themselves, as particle aggregates form over time. Fish might chase these aggregates, which could be considered a feeding behavior [38]. Consequently, the potential effects of GO may be linked to particle size, exposure duration, and the underlying processes of the harmful action of these chemicals on specific target organs. These effects depend on the fish's stage of development and may result in diverse biological responses, including those at a whole-organismal level. Despite numerous approaches to explaining alterations in the behavior of fish embryos and larvae attributed to the mode of action of graphene-based nanomaterials (GBNs), this area of research is still ongoing.

Existing limitations in GBN extraction from water and sediments impede precise assessments in aquatic environments. For a thorough prediction of GBNs' adverse effects, future studies should consider nanomaterial type, characteristics, water behavior, potential concentrations, and various environmental abiotic factors influencing aquatic life. Analysis of the literature suggests environmental concentrations of various carbon nanomaterials may range from 0.1 up to 1000 μg L^{-1} [17,39], serving as an initial point for nanocarbon ecotoxicological analysis. As nanomaterial production advances across scientific domains, concentrations are likely to increase, underscoring the necessity for comprehensive studies on GBNs' toxicity across diverse environmental contexts for both aquatic and terrestrial species.

5. Conclusions

This study examined the effects of two common carbon nanomaterials, graphene (nanoflakes) and its oxidized form, graphene oxide (GO), on early life salmonids (*S. trutta*, *O. mykiss*). The investigation explored both short- (up to 96 h) and long-term (up to 23 days) impacts on fish nesting behavior, combined with cardiorespiratory responses, and swimming performance. The results revealed that a short-term exposure to GO significantly influenced the behavioral responses of *S. trutta* larvae, causing an 11–23% reduction in nesting activity and a 54–72% increase in swimming activity among fish juveniles. Examination of cardiorespiratory endpoints in *S. trutta* juveniles showed a 9% decrease in gill ventilation frequency of larvae indicating physiological changes. Notably, graphene (nanoflakes) had a prolonged impact on *O. mykiss* larvae, leading to an increased level of fish swimming activity over a 20-day exposure period. The study concluded that GO exhibits concentration-dependent effects on the studied fish.

This study shows that studied carbon nanomaterials significantly affect fish nesting and swimming behavior in early development, potentially impacting population dynamics and individual survival in their natural habitats. Additional research is essential to grasp the safety profile of various carbon nanomaterials and their potential toxicity for diverse organisms. This includes a broader examination of biological responses in conjunction with the dynamic nature of the environment.

Author Contributions: Conceptualization, T.M. and M.J.-L.; formal analysis, T.M. and Ž.J.; methodology, T.M., Ž.J., S.Š. and M.J.-L.; investigation, T.M., Ž.J., S.Š. and M.J.-L.; data curation, T.M.; writing—original draft preparation, T.M.; visualization, T.M. and M.J.-L.; writing—review and editing, T.M., Ž.J., S.Š. and M.J.-L.; project administration, T.M. and M.J.-L. All authors have read and agreed to the published version of the manuscript.

Funding: This research was funded by statutory projects DOT20/STRES and DOT21/STRES conducted at the National Marine Fisheries Research Institute and financed by the Ministry of Science and Higher Education in Poland and by the Research Council of Lithuania, Project No. S-MIP-22-51.

Institutional Review Board Statement: The animal study protocol was approved by the Institutional Review Board of Lithuanian State Food Veterinary Service (license no. G2-168; valid from 8 February 2021 to 7 February 2025) for studies involving animals.

Informed Consent Statement: Not applicable.

Data Availability Statement: The data presented in this study are available on request from the corresponding author.

Conflicts of Interest: The authors declare no conflicts of interest.

References

1. Singh, S.; Hasan, M.R.; Sharma, P.; Narang, J. Graphene nanomaterials: The wondering material from synthesis to applications. *Sens. Int.* **2022**, *3*, 100190. [CrossRef]
2. Pan, X.; Ji, J.; Zhang, N.; Xing, M. Research progress of graphene-based nanomaterials for the environmental remediation. *Chin. Chem. Lett.* **2020**, *31*, 1462–1473. [CrossRef]
3. Kumar, V.; Kim, K.H.; Park, J.W.; Hong, J.; Kumar, S. Graphene and its nanocomposites as a platform for environmental applications. *Chem. Eng. J.* **2017**, *315*, 210–232. [CrossRef]
4. De Marchi, L.; Pretti, C.; Gabriel, B.; Marques, P.A.A.P.; Freitas, R.; Neto, V. An overview of graphene materials: Properties, applications and toxicity on aquatic environments. *Sci. Total Environ.* **2018**, *631–632*, 1440–1456. [CrossRef] [PubMed]
5. Borandeh, S.; Alimardani, V.; Abolmaali, S.S.; Seppälä, J. Graphene Family Nanomaterials in Ocular Applications: Physicochemical Properties and Toxicity. *Chem. Res. Toxicol.* **2021**, *34*, 1386–1402. [CrossRef] [PubMed]
6. Kumar, S.; Parekh, S.H. Linking graphene-based material physicochemical properties with molecular adsorption, structure and cell fate. *Commun. Chem.* **2020**, *3*, 8. [CrossRef] [PubMed]
7. Ou, L.; Song, B.; Liang, H.; Liu, J.; Feng, X.; Deng, B.; Sun, T.; Shao, L. Toxicity of graphene-family nanoparticles: A general review of the origins and mechanisms. *Part. Fibre Toxicol.* **2016**, *13*, 57. [CrossRef] [PubMed]
8. Liao, C.; Li, Y.; Tjong, S.C. Graphene Nanomaterials: Synthesis, Biocompatibility, and Cytotoxicity. *Int. J. Mol. Sci.* **2018**, *19*, 3564. [CrossRef] [PubMed]

9. Li, J.; Zeng, H.; Zeng, Z.; Zeng, Y.; Xie, T. Promising Graphene-Based Nanomaterials and Their Biomedical Applications and Potential Risks: A Comprehensive Review. *ACS Biomater. Sci. Eng.* **2021**, *7*, 5363–5396. [CrossRef]
10. Malhotra, N.; Villaflores, O.B.; Audira, G.; Siregar, P.; Lee, J.S.; Ger, T.R.; Hsiao, C.D. Toxicity Studies on Graphene-Based Nanomaterials in Aquatic Organisms: Current Understanding. *Molecules* **2020**, *25*, 3618. [CrossRef]
11. Ding, X.; Pu, Y.; Tang, M.; Zhang, T. Environmental and health effects of graphene-family nanomaterials: Potential release pathways, transformation, environmental fate and health risks. *Nanotoday* **2022**, *42*, 101379. [CrossRef]
12. Chen, M.; Yin, J.; Liang, Y.; Yuan, S.; Wang, F.; Song, M.; Wang, H. Oxidative stress and immunotoxicity induced by graphene oxide in zebrafish. *Aquat. Toxicol.* **2016**, *174*, 54–60. [CrossRef]
13. Souza, J.P.; Baretta, J.F.; Santos, F.; Paino, I.M.M. Toxicological effects of graphene oxide on adult zebrafish (*Danio rerio*). *Aquat. Toxicol.* **2018**, *186*, 11–18. [CrossRef]
14. Dasmahapatra, A.K.; Dasari, T.P.S.; Tchounwou, P.B. Graphene-Based Nanomaterials Toxicity in Fish. *Rev. Environ. Contam. Toxicol.* **2019**, *247*, 1–58.
15. Chen, Z.; Yu, C.; Khan, I.A.; Tang, Y.; Liu, S.; Yang, M. Toxic effects of different-sized graphene oxide particles on zebrafish embryonic development. *Ecotoxico. Environ. Saf.* **2020**, *197*, 110608. [CrossRef]
16. Kalman, J.; Merino, C.; Fernández-Cruzm, M.L.; Navas, J.M. Usefulness of fish cell lines for the initial characterization of toxicity and cellular fate of graphene-related materials (carbon nanofibers and graphene oxide). *Chemosphere* **2019**, *218*, 347–358. [CrossRef]
17. Zhang, X.; Zhou, Q.; Zou, W.; Hu, X. Molecular Mechanisms of Developmental Toxicity Induced by Graphene Oxide at Predicted Environmental Concentrations. *Environ. Sci. Technol.* **2017**, *51*, 7861–7871. [CrossRef] [PubMed]
18. Mauro, G.; Rauti, R.; Casani, R.; Chimowa, G.; Galibert, A.M.; Flahaut, E.; Cellot, G.; Ballerini, L. Tuning the reduction of graphene oxide nanoflakes differently affects neuronal networks in the zebrafish. *Nanomaterials* **2021**, *11*, 2161. [CrossRef]
19. Cao, Z.; Su, M.; Wang, H.; Zhou, L.; Meng, Z.; Xiong, G.; Liao, X.; Lu, H. Carboxyl graphene oxide nanoparticles induce neurodevelopmental defects and locomotor disorders in zebrafish larvae. *Chemosphere* **2021**, *270*, 128611. [CrossRef]
20. Dasmahapatra, A.K.; Powe, D.K.; Dasari, T.P.S.; Tchounwou, P.B. Assessment of reproductive and developmental effects of graphene oxide on Japanese medaka (*Oryzias latipes*). *Chemopshere* **2020**, *259*, 127221. [CrossRef]
21. Dasmahapatra, A.K.; Tchounwou, P.B. Histopathological evaluation of the interrenal gland (adrenal homolog) of Japanese medaka (*Oryzias latipes*) exposed to graphene oxide. *Environ. Toxicol.* **2022**, *37*, 2460–2482. [CrossRef] [PubMed]
22. Jakubowska, M.; Białowąs, M.; Stankevičiūtė, M.; Chomiczewska, A.; Jonko-Sobuś, K.; Pažusienė, J.; Hallmann, A.; Bučaitė, A.; Urban-Malinga, B. Effects of different types of primary microplastics on early life stages of rainbow trout (*Oncorhynchus mykiss*). *Sci. Total Environ.* **2022**, *808*, 151909. [CrossRef]
23. Directive. 63/EU of the European Parliament and of the Council of 22nd September 2010 on the protection of animals used for scientific purposes. *Off. J. Eur. Union* **2010**, *276*, 33–79. Available online: https://eur-lex.europa.eu/LexUriServ/LexUriServ.do?uri=OJ:L:2010:276:0033:0079:en:PDF (accessed on 9 April 2024).
24. Makaras, T.; Jurgelėnė, Ž.; Pakštas, V.; Šemčuk, S. Interacting effects of graphene oxide derivatives and trace metals on freshwater brown trout (*Salmo trutta* L.) behaviour at different stages of development. *ASC Environ. Sci. Nano* **2023**, *10*, 2172–2184. [CrossRef]
25. Claramunt, S.; Varea, A.; López-Díaz, D.; Velázquez, M.M.; Cornet, A.; Cirera, A. The importance of interbands on the interpretation of the Raman spectrum of graphene oxide. *J. Phys. Chem. C* **2015**, *119*, 18. [CrossRef]
26. Krishnamoorthy, K.; Veerapandian, M.; Yun, K.; Kim, S.J. The chemical and structural analysis of graphene oxide with different degrees of oxidation. *Carbon* **2013**, *53*, 38–49. [CrossRef]
27. Jakubowska-Lehrmann, M.; Dąbrowska, A.; Białowąs, M.; Makaras, T.; Hallmann, A.; Urban-Malinga, B. The impact of various carbon nanomaterials on the morphological, behavioural, and biochemical parameters of rainbow trout in the early life stages. *Aquat. Toxicol.* **2023**, *259*, 106550. [CrossRef]
28. *ISO 7346-1:1996*; Determination of the Acute Lethal Toxicity of Substances to a Freshwater Fish [Brachydanio rerio Hamilton-Buchanan (Teleostei, Cyprinidae)]—Part 1: Static Method. ISO: Geneva, Switzerland, 1996.
29. OECD. Section 2 Effects on Biotic Systems, Test No. 210: Fish, Early-life Stage Toxicity Test. In *OECD Guidelines for the Testing of Chemicals*; OECD Publishing: Paris, France, 2013. [CrossRef]
30. Soares, J.C.; Pereira, T.C.B.; Costa, K.M.; Maraschin, T.; Basso, N.R.; Bogo, M.R. Developmental neurotoxic effects of graphene oxide exposure in zebrafish larvae (*Danio rerio*). *Colloids Surf. B Biointerfaces* **2017**, *157*, 335–346. [CrossRef]
31. Liu, X.T.; Mu, X.Y.; Wu, X.L.; Meng, L.X.; Guan, W.B.; Qiang, Y.; Li, X.F. Toxicity of multi-walled carbon nanotubes, graphene oxide, and reduced graphene oxide to zebrafish embryos. *Biomed. Environ. Sci.* **2014**, *27*, 676–683.
32. Chinnadurai, G.; Subramanian, R.; Maqusood, A. Fish mucus mediated biosynthesis of copper oxide nanoparticles: Spectral characterization, morphology and biological activity. *Mater. Res. Express* **2020**, *7*, 125012. [CrossRef]
33. Farrell, A.P. Cardiorespiratory performance during prolonged swimming tests with salmonids: A perspective on temperature effects and potential analytical pitfalls. *Philos. Trans. R. Soc. B* **2007**, *362*, 2017–2030. [CrossRef]
34. Clemente, Z.; Silva, G.H.; de Souza Nunes, M.C.; Martinez, D.S.T.; Maurer-Morelli, C.V.; Thomaz, A.A.; Castro, V.L.S.S. Exploring the mechanisms of graphene oxide behavioural and morphological changes in zebrafish. *Environ. Sci. Pollut. Res.* **2019**, *26*, 30508–30523. [CrossRef] [PubMed]
35. Wang, Z.G.; Rong, Z.H.O.U.; Jiang, D.; Jing, E.S.; Qian, X.U.; Jing, S.I.; Bin, L. Toxicity of graphene quantum dots in zebrafish embryo. *Biomed. Environm. Sci.* **2015**, *28*, 341–351.

36. Yang, X.; Yang, Q.; Zheng, G.; Han, S.; Zhao, F.; Hu, Q. Developmental neurotoxicity and immunotoxicity induced by graphene oxide in zebrafish embryos. *Environ. Toxicol.* **2019**, *34*, 415–423. [CrossRef]
37. Handy, R.D.; Henry, T.B.; Scon, T.M.; Johnston, B.D.; Tyler, C.R. Manufactured nanoparticles: Their uptake and effects on fish d a mechanistic analysis. *Ecotoxicology* **2008**, *17*, 396–409. [CrossRef]
38. Andrew, J.E.; Holm, J.; Kadri, S.; Huntingford, F.A. The effect of competition on the feeding efficiency and feed handling behaviour in gilthead sea bream (*Sparus aurata* L.) held in tanks. *Aquaculture* **2004**, *232*, 317–331. [CrossRef]
39. Goodwin, D.G., Jr.; Adeleye, A.S.; Sung, L.; Ho, K.T.; Burgess, R.M.; Petersen, E.J. Detection and quantification of graphene-family nanomaterials in the environment. *Environ. Sci. Technol.* **2018**, *52*, 4491–4513. [CrossRef]

Journal of *Xenobiotics*

Article

Bioaccumulation of Organic and Inorganic Pollutants in Fish from Thermaikos Gulf: Preliminary Human Health Risk Assessment Assisted by a Computational Approach

Konstantinos M. Kasiotis [1,*], Effrosyni Zafeiraki [1,*], Electra Manea-Karga [1], Demetrios Kouretas [2], Fotis Tekos [2], Zoi Skaperda [2], Nikolaos Doumpas [3] and Kyriaki Machera [1]

1 Laboratory of Pesticides' Toxicology, Department of Pesticides Control and Phytopharmacy, Benaki Phytopathological Institute, 8 Stefanou Delta Str., 14561 Athens, Greece; e.manea-karga@bpi.gr (E.M.-K.); k.machera@bpi.gr (K.M.)

2 Department of Biochemistry and Biotechnology, University of Thessaly, Viopolis, Mezourlo, 41500 Larissa, Greece; dkouret@uth.gr (D.K.); fotis.tek@gmail.com (F.T.); skaperda@foodoxys.com (Z.S.)

3 iSea, Environmental Organisation for the Preservation of the Aquatic Ecosystems, 54645 Thessaloniki, Greece; nikolaos.doumpas@isea.com.gr

* Correspondence: k.kasiotis@bpi.gr (K.M.K.); e.zafeiraki@bpi.gr (E.Z.); Tel.: +30-2108180357 (K.M.K. & E.Z.)

Abstract: The monitoring of contaminants in fish species is pivotal for fishes' health and reproduction, as well as for human health. In the specific work, three major categories of contaminants, pesticides, pharmaceuticals, and macro and trace elements, were investigated in two major fish species, *Dicentrarchus labrax* and *Solea solea*, collected from Thermaikos Gulf, in Greece. To achieve this goal, three analytical methods using LC-MS/MS, GC-MS/MS, and ICP-MS were developed, validated, and applied to the collected fish samples. The results indicated a very low prevalence of caffeine and acetaminophen, both not exceeding 3.8 μg/kg fish. Similarly, thiabendazole, cypermethrin, and tricyclazole (pesticides) were found in a concentration range of 0.9 to 13.7 μg/kg fish, while in one *D. labrax* sample, traces of the metabolite of organochlorine pesticide DDT, *o,p*′-DDE were detected. Al, Mn, Fe, Zn, and Sr were the predominant trace elements in a concentration range of 500–20,000 μg/kg fish. Macro elements levels varied from 280 to 5405 mg/kg fish. Health risk assessment did not unveil an unacceptable risk for the human health of adults, apart from one sample presenting Hg above the regulatory levels. On the contrary, for children, the calculated hazard quotient values for Hg in all cases and for two As detections were higher than the threshold value of 1, indicating a potential risk.

Keywords: fish; pesticides; pharmaceuticals; heavy metals; Mediterranean; liquid chromatography; gas chromatography; inductive coupled plasma; mass spectrometry

1. Introduction

Marine pollution has gained scientific interest over the last decades, as the presence of pollutants agitates the natural balance of the ecosystems and may pose a threat to the environment and to human health. Coastal areas, including estuarine, bays, and gulfs, are considered to be more vulnerable to the accumulation of pollutants, since they are heavily impacted by both natural and anthropogenic activities [1]. The European Union's Marine Strategy Framework Directive (MSFD) has set a clear framework for the maintenance of proper marine environmental status, embracing quality descriptors such as biodiversity status and contaminant levels in fish [2]. In this context, the investigation of chemicals (such as pharmaceuticals) in the marine environment has long been an objective of many studies and was reviewed recently [3], with compounds such as acetaminophen and ibuprofen holding a main position among the analgesic compounds and non-steroidal anti-inflammatory drugs, respectively, frequently found in seawater, including the Mediterranean Sea, with relatively high concentrations [4,5]. Similarly, heavy metals and pesticides are also at the forefront, with the described undesired health effects on fish commodities and aquatic organisms in broader sense [6–10] included in marine monitoring schemes [11–13].

Thermaikos Gulf is a semi-closed shallow gulf in the northwestern Aegean Sea, characterized by intense activities taking place in its coastal area. Thermaikos Gulf is the recipient of the input of four rivers (Axios, Aliakmon, Gallikos and Loudias) and two municipal wastewater treatment plants of Thessaloniki. Additional pressure received by the discharges of industrial, agricultural, marine, and aquaculture activities occurring at the coastline of the gulf, makes the latter a reservoir of pollutants. The elevated chemical burden in the marine and coastal environment of the Gulf yields effects on both marine ecosystem and humans [14]. Since fishing and mussel farming are economic drivers for the coastal cities of Thermaikos (from Thessaloniki to Larissa) [15], the required balance between such activities and ecological and socioeconomic targets should not be disregarded, and is usually safeguarded by marine spatial plans [16]. Considering that Thermaikos Gulf is one of the major fishing grounds of Greece and this part of Mediterranean [14,17–19], the occurrence of potentially toxic chemicals in aquatic organisms can provoke great health risks to the consumers.

According to the literature, many undesirable substances, such as pesticides, pharmaceuticals, and heavy metals, are absorbed and finally accumulate in the marine organisms [20,21]. Due to the toxicity of these pollutants, their presence in fish is of critical significance. However, the available information on organic and inorganic pollutants in fish from Thermaikos Gulf is still limited, with most of the studies focusing on this area providing data related to the sediment, algae, and water pollution [22–27].

Identifying the need for control strategies and continuous monitoring of pollutants in aquatic organisms in Thermaikos Gulf, in the present study, we elaborate the detection of several organic and inorganic pollutants in the euraline fish species, *Dicentrarchus labrax* (*D. labrax*, European seabass) and *Solea solea* (*S. solea*, common sole). In particular, 286 chemicals, including pesticides, pharmaceuticals, trace, and macro elements, were incorporated into the analytical methods to investigate the occurrence of the former and further evaluate the impact of the pressure upon the Thermaikos ecosystem. The potential risk to human health posed by the most hazardous compounds or elements detected was also assessed and discussed. Overall, the herein concurrent application of several multi-analyte methods covering a broad range of chemicals in marine organisms is of high significance due to the global need for the generation of contaminants data that, to an extent, portray the contaminants' burden on the sampled area and in parallel feed health risk assessments.

2. Materials and Methods

2.1. Samples

Fish samples of *D. labrax* and *S. solea* [28] were collected from the internal part of Thermaikos Gulf (in proximity to the coastal line of Thessaloniki, see Figure 1) in the areas of Perea, Aggelochori, and Epanomi during spring 2021 following a well-established protocol [29] and were placed into distinct resealable bags with a label declaring pertinent information (e.g., date, time, site, species). Since samples treatment/dissection was not possible within 24 h, samples were frozen at −20 °C. Prior to chemical analysis, samples were unfrozen, and the whole fish was homogenized.

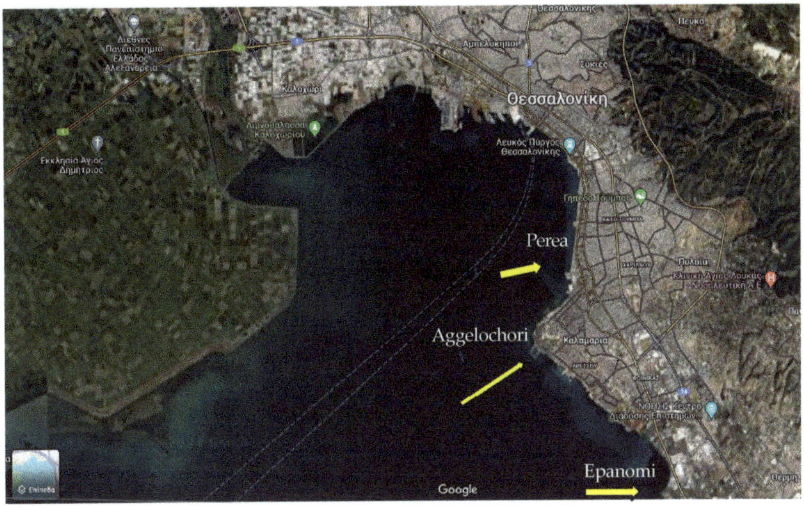

Figure 1. Map of sampling points (photo retrieved from Google Maps).

2.2. Chemicals, Materials, and Standard Solutions

Some of the chemicals used for the analysis of pesticide residues were previously reported in other published works from our group [30–33]. Enhanced matrix removal (EMR)-lipid clean-up dispersive SPE (dSPE) was purchased from Agilent Technologies. The mass-labeled internal standards, carbendazim D3, imidacloprid-D4, dimethoate-D6, and chlropyrifos-D10 were used for the quantification of the compounds measured with LC ESI-MS/MS, and they were purchased from Sigma Aldrich (Seelze, Germany). Triphenyl phosphate (TPP) and dichlorvos-D6 were used in the GC-MS/MS analysis, and were acquired from Sigma-Aldrich (Seelze, Germany). The overall analytical scope covered approximately 250 pesticides (see Table S1 for the analytes included). In total, 4 Antibiotics and 6 pharmaceuticals were selected to be monitored, taking into consideration the prevalence of such compounds in wastewater and the marine environment in general, including the Mediterranean Sea [4,11,34]. Chemicals included were acetaminophen, amoxicillin, azithromycin, atorvastatin, caffeine, cefuroxime, clarithromycin, ibuprofen, ramipril, and venlafaxine. All analytical standards were obtained from Sigma Aldrich (Seelze, Germany) and were of a purity >99%. Distinct stock solutions at 1000 µg/mL were prepared in organic solvent (MeOH for the LC-ESI-MS/MS, and acetone or acetonitrile for GC-MS/MS), and intermediate/working solutions were prepared by appropriate dilutions in respective solvents.

Twenty seven (27) elements were determined and quantified by using a standard mix of 25 chemical elements (Al, Ba, B, Cu, Fe, Sr, Zn, Be, Cr, Co, Li, Mn, Mo, Ni, Ti, V, Sb, As, Cd, Pb, Se, Ag, Tl, Sn, and U in HNO_3 5%) and standards solutions of individual elements (Hg in 10% HNO_3, P in 0,05% H_2SO_4, Mg, and Ca in 2% HNO_3), purchased from CPAchem (Stara Zagora, Bulgaria). Internal standards: individual solutions of Lithium (6Li), Scandium (Sc), Germanium (Ge), Ytrium (Y), Indium (In), Terbium (Tb), and Iridium (Ir), and a standard solution/mixture of 25 element components (Al, Be, Co, Li, Se, Sn, Zn, Sb, B, Cu, Mn, Ag, Ti, As, Cd, Fe, Mo, Sr, U, Ba, Cr, Pb, Ni, Tl, and V in HNO_3 5%) were used for the preparation of quality assurance (QC) standard and were also provided by CPAchem (Stara Zagora, Bulgaria). The European Reference Material, ERM®-BB422 fish muscle, was obtained from the Joint Research Centre (JRC) of the European Commission (Certified Reference Material (CRM), fish muscle fortified with 14 elements at known concentration, JRC, Institute for Reference Materials and Measurements, Geel, Belgium). Speciation analysis for As and Cr (both organic and inorganic forms) was not performed during this study. Nitric acid (HNO_3) 67–69% w/w and hydrogen peroxide (H_2O_2) 30%

w/w (trace metal grade) were all purchased from Seastar Chemicals Inc. (Sidney, Canada). Ultrapure water from an ultrapure Milli-Q water system (Burlington, MA, USA) was also used for the dilution of all the solutions.

2.3. Sample Preparation

2.3.1. Pesticide Residues and Pharmaceuticals Analysis

Blank fish samples of both fish types (of organic fish farming) were obtained from the market, since, to our knowledge, a CRM-negative fish tissue was not available (taking into account all classes of chemicals screened in this work). The samples were homogenized in their totality and used to validate the analytical methods. Prior to the validation of the methods, the samples underwent sample preparation to verify the absence of the chemicals studied herein. Since no significant difference in terms of matrix interferences between the two types of fish was obtained, only one was used for thorough validation trials.

The sample preparation for the determination and quantification of pesticides and pharmaceuticals was based on a previous work from our team [30], applying some modifications proposed in the literature concerning the more efficient removal of lipids in lipid-rich matrices [35,36]. More specifically, homogenized fish (2 g whole fish) were placed inside a falcon tube (50 mL) and spiked with six internal standards, carbendazim-D3, imidachloprid-D4, dimethoate-D6, dichlorvos-D6, chlorpyriphos-D10, and TPP at a concentration of 50 ng/g. Then, ACN (10 mL), $MgSO_4$ (2 g), and 1 g sodium acetate were introduced sequentially, and the tube was vigorously hand-shaken for 1 min and vortex-mixed for 3 min to achieve homogenization. Consequently, the mixture was centrifuged for 10 min at 4500 rpm (Hettich Zentrifugen, UNIVERSAL 32R, Tuttlingen, Germany) with a stable temperature of 10 °C. The upper layer was transferred to another falcon tube (15 mL), where EMR lipid agent (500 mg) and 3 mL H_2O were placed. The mixture was vortex-mixed for 1.5 min and centrifuged for 5 min at 4500 rpm and 10 °C. The supernatant was divided into two parts, and both aliquots were evaporated to dryness using a N_2 stream at 35 °C. Consequently, the dried extracts were reconstituted with 1 mL of a 75:25 (v/v) MeOH:H_2O and 1 mL of pure acetone and injected into the LC-MS/MS and GC-MS/MS system, respectively (before injection filtering through a PTFE filter (0.22 μm) was performed). Considering the acceptable recoveries obtained during method validation and to reduce the analysis cost, additional internal standards of the pharmaceuticals were not acquired.

2.3.2. Trace and Macro Elements Analysis

For the analysis of trace and macro elements in fish samples, 0.5 g of every homogenized fish sample was weighed and placed into a microwave oven vessel (ETHOS UP, Milestone) for digestion. HNO_3 (9 mL) and H_2O_2 (1 mL) were added in each vessel, and the samples were further digested under 210 °C and 1800 W for 35 min. The vessels were then cooled to less than 40 °C, and the digestion solution was transferred to a clean container. Ultrapure water was added until a final volume of 100 mL. All the labware and the vessels were acid-washed (immersion into a diluted HNO_3 0.05% with ultrapure water) for 24 h and then washed several times with ultrapure water. Then, they were dried in a clean hood at room temperature.

2.4. Instrumental Analysis

2.4.1. Liquid Chromatography Electron Spray Ionization Tandem Mass Spectrometry (LC-ESI-MS/MS)

An Agilent triple quadrupole (6410B, QQQ) system was used for the analysis of the samples and the detection of pesticide residues. The LC was equipped with a C18 column, and the chromatographic gradient started from 80% mobile phase A (% 5 mM ammonium formate in H_2O, 0.1% formic acid) to 100% mobile phase B (5 mM ammonium formate in MeOH) in 30 min. An injection volume of 20 μL was applied, while the flow rate was set at 0.28 mL/min. Results were also corroborated in an additional triple quadrupole system, namely a triple Quad LC/MS system (Shimadzu LCMS-8060 NX, Kyoto, Japan), using the same conditions described above, considering the enhanced performance of this system in terms of sensitivity for some of the analytes (i.e., tricyclazole and ibuprofen).

2.4.2. Gas Chromatography Tandem Mass Spectrometry (GC-MS/MS)

The GC-MS/MS analysis of the samples was performed on a Shimadzu triple quadrupole mass spectrometer GCMS-TQ8040NX composed of a GC-2030 gas chromatograph. The GC system was equipped with an autosampler AOC-6000 (CTC Analytics) and a 10 µL syringe. The capillary column used for the chromatographic separation was the MEGA-5 HT (30 m × 0.25 mm × 0.25 µm) (MEGA S.r.l., Legnano, Italy), and the chromatographic gradient applied was the following: initial oven temperature of 50 °C for 1 min, increase in T until 125 °C at a rate of 25 °C/min, and finally reach of 300 °C at a rate of 10 °C/min. The analyses in MS/MS were performed in multiple reaction monitoring (MRM) mode.

2.4.3. Inductive Coupled Plasma Mass Spectrometry (ICP-MS)

The ICP-MS analysis was performed on a Thermo iCAP-RQ equipped with an ASX-280 autosampler, as previously described in detail [37]. Briefly, the instrument used a Ni sample and skimmer cones, a Quartz cyclonic spray chamber, and a MicroMist U-Series Nebulizer (0.4 mL/min with PEEK connector). The flow of the nebulizer gas and the cool gas was approximately 1 L/min (autotune dependent) and 14 L/min, respectively. Prior to the analysis, the system was equilibrated, and a performance report in KED mode was performed. In this context, the analysis was conducted with high sensitivity and stability, while the occurrence of interferences and doubly charged ions was minimized.

2.5. *Quantification and Quality Assurance*

2.5.1. Pesticide Residues

The two multiresidue methods of LC-ESI-MS/MS and GC-MS/MS were validated based on the SANTE Guidance Document on analytical quality control and method validation procedures for pesticide residues analysis in food and feed [38]. The method limit of quantification (LOQ) for pesticides was established as the lowest fortification level that provided sufficient accuracy and precision. The LOQ and recovery values for each active substance are presented in Table S1. Three control samples were analyzed in every sequence of samples, and no pesticide residues or their metabolites were detected in any of them. All validation metrics, such as selectivity, working concentration range and linearity, LOQ, precision, and accuracy, were determined following the steps described in the SANTE guidance document.

2.5.2. Pharmaceuticals

For pharmaceuticals, method validation was also based on the SANTE document [38]. Yet, the limits of detection LODs and LOQs were estimated from the analyte signals observed after LC-ESI-MS/MS analysis of fortified fish samples, as the concentration that provided a signal-to-noise ratio of 3 in the case of the LOD and 10 in the case of the LOQ (presented in Table S2). The linearity was evaluated through seven calibration points, namely 1, 5, 10, 20, 50, 100, and 200 ng/mL, and considered acceptable when the regression coefficient (R^2) was above 0.99. MRM conditions, including collision energies applied and RT for each pharmaceutical, are illustrated in Table S3. For both pharmaceuticals and pesticides, quantification was conducted with external standard calibration using matrix-matched calibration curves.

2.5.3. Trace and Macro Elements

Solvent calibration curves covering concentrations from 0.1 to 1000 µg/Kg and from 0.1 to 200 mg/Kg were prepared for the quantification of trace elements (Al, Ti, V, Cr, Mn, Fe, Co, Ni, Cu, Zn, As, Se, Sr, Mo, Ag, Cd, Sn, Sb, Ba, Tl, Pb, Hg, and U) and macro elements (Mg, P, K, and Ca), respectively. The regression coefficient (r^2) was greater than

0.995 for all the calibration curves, while seven internal standards (^{6}Li, Sc, Ge, Y, In, Tb, and Ir) were added to all unknown samples and calibration standards at a constant rate. The limits of quantification (LOQ) values and the quantification parameters for each element are presented in Table S4. Two quality control (QC) standards solutions of macro and trace elements were analyzed in every batch of samples, monitoring the repeatability of the analytical method. The recovery of the two QC standards ranged between 80–120% in all the samples. In addition, blank reagents, spiked samples, and ERM were analyzed following the same procedure as the unknown samples and measured in every sequence to control the background contamination and the accuracy of the method. Moreover, one additional isotope was measured when possible to avoid spectral interferences. To compare the metal concentrations between the two fish species, a one-way analysis of variance (ANOVA) was conducted. The two fish species were set up as the independent variable and the metal concentration was set up as the dependent variable. This allowed us to evaluate whether there were significant differences in metal concentrations between the two species.

2.6. Health Risk Assessment

The risk on human health posed by each compound and element detected in the analyzed fish samples was assessed based on the Hazard Quotient (HQ) and Hazard Index (HI) approach [39]. For the calculation of the HQ, the following equations were applied:

$$HQ = \frac{ADD}{TDI} \quad (1)$$

$$ADD = C \times \frac{IR}{BW} \quad (2)$$

where:

ADD: average daily intake (μg/Kg bw/d).

TDI: the tolerable daily intake (μg/Kg bw/d) set by European Food Safety Authority (EFSA) and open literature.

C: the mean of each compound or element concentration detected in fish (μg/Kg).

IR: the daily fish consumption rate (Kg/person/d) for Greece (fish: 0.056 Kg/inhabitant/day, [40]).

BW: the mean body weight (70 Kg for adults and 15 Kg for children).

For the estimation of the cumulative health risk, the Hazard Index approach (HI, unitless) was applied (Equation (3)). More specifically, the hypothesis of dose additivity was calculated by summarizing the individual HQ values.

$$HI = \sum HQs = HQ1 + HQ2 + HQ3 + \cdots + HQn \quad (3)$$

2.7. In Silico Calculations

Nexus software (version 2.5.1, Lhasa Limited) and Derek Nexus software (version 6.2.0, Lhasa Limited) were used for the in silico approach to corroborate the findings on pharmaceuticals and their relation to health risk assessment. More specifically, more than 60 endpoints were selected, including carcinogenicity, cardiotoxicity, developmental toxicity, hepatotoxicity, neurotoxicity, and teratogenicity. Concerning species, both mammals and bacterium were regarded. For each compound, structural details exemplified by LogP, LogKp, exact, and average molecular mass were considered, while tautomers and mixtures (for the same compound) were also deemed in computations. The Derek KB 2022 1.0 knowledge base was used to retrieve toxicological information.

3. Results

3.1. Analytical Methods Validation

The fortification experiments resulted in acceptable recoveries for all analytes in the range of 64–117% (see Table S1 for pesticides and Table S2 for pharmaceuticals). Linearity was acceptable with regression coefficients (R^2) values >0.99 (with the exception of paclobutazole and diphenylamine, where R^2 were 0.9889 and 0.9881, respectively) accompanied by acceptable residual values. Since a CRM-negative fish tissue sample was not obtained (see Section 2.3.1), focus was given to the assessment of blanks used in this study. The blank extracts were injected into the analytical instruments, and the analytical signal of all analytes was assessed. In limited cases (carbendazim, cyfluthrin, hexaconazole, fenazaquin, oxamyl, mepanipyrim, malathion, fenthion, pymethrozine, pyrazophos, and ramipril), detection of analytes in blank samples was in the range of <LOD to LOQ/4. Therefore, the subtraction of analyte concentration in blank extracts was not deemed necessary. Lastly, the measured concentrations of elements in ERM®-BB422 fish muscle were in agreement with the assigned values.

3.2. Levels of Pesticide Residues in Fish Samples from Thermaikos Gulf

Eight out of ten samples were devoid of pesticide residues. Surprisingly, in one *S. solea* sample, thiabendazole was quantified at 13.7 μg/Kg fish, which is considered a relatively high concentration to be detected in fish collected from the open sea. Thiabendazole has been reported in South African freshwater impoundment [41], although in the same study, thiabendazole residues were not obtained in fish tissues. Similarly, in Spain, thiabendazole was reported as being among the compounds with the highest concentrations in river water, although not among the most frequently detected [42]. Another azole-type molecule, tricyclazole, was detected in the same sample, but at a lower concentration (at 0.9 μg/Kg fish). On the other hand, cypermethrin was quantified in one sample of *D. labrax* at 8.7 μg/Kg fish (see Figure 2). To our knowledge, for the abovementioned compounds, no MRLs are established in fish commodities by the European Union [43]. Expanding the comparison of findings with the Codex Alimentarius International Food Standard, US-EPA tolerances, and Australian MRLs, similar conclusions were derived, since respective limits in fish are also not established [44–46]. Traces of o,p'-DDE (in proximity to the LOD of the method at 0.15 μg/Kg) were detected in one *D. labrax* sample (see Figure S1c).

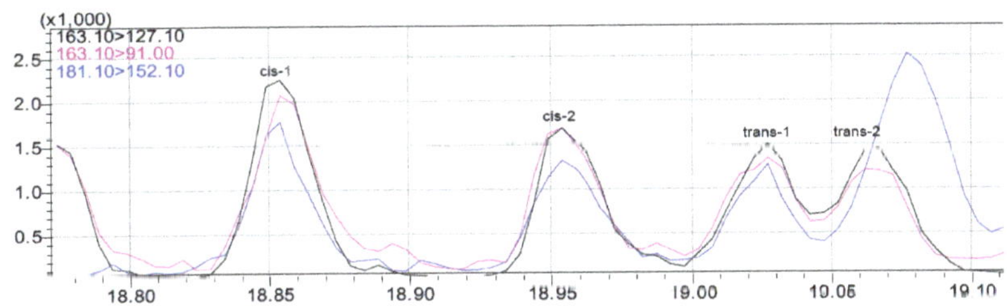

Figure 2. MRM chromatograms of cypermethrin (as a sum of cis and trans diastereomers) in a positive sample.

3.3. Levels of Pharmaceuticals in Fish Samples from Thermaikos Gulf

Low prevalence of the selected pharmaceuticals was obtained in the specific work. More specifically, the most prevalent pharmaceutical was caffeine, detected in three samples (two *D. labrax* samples and one *S. solea* sample; for indicative chromatogram, see Figure 3, while for blank and spiked samples, see Figure S2a and b, respectively) in the range of 2.3–3.8 μg/Kg fish. In the same context, acetaminophen was quantified in one sample (*S. solea*) at 1.5 μg/Kg fish. Concomitant detections of pharmaceuticals and pesticides occurred only in one sample (cypermethrin and caffeine in the *D. labrax* sample).

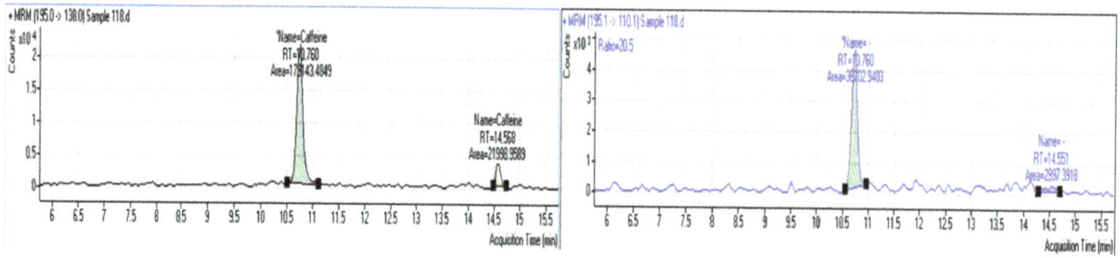

Figure 3. MRM chromatograms of caffeine in a positive sample.

For both caffeine and acetaminophen, no MRL was established in foodstuffs of animal origin. Traces of ibuprofen (at the LOD) were detected in one *D. labrax* sample (see Figure S2c).

3.4. Levels of Trace and Macro Elements in Fish Samples from Thermaikos Gulf

The accumulation of 27 trace and macro elements was investigated in two different species of fish collected from Thermaikos Gulf. The detected concentrations for each individual element are presented in Tables S4 and S5, while the skewness of the most abundant trace elements and of macro elements for each species is also illustrated in Figures 4 and 5, respectively, by displaying the data averages and quartiles. According to the results, trace and macro elements were detected in all the fish samples (n = 10). The highest detected concentrations were those of macro elements (Mg, Ca, P, and K), followed by the trace elements Sr, Fe, Zn, Al, and Mn. In particular, the average concentrations of macro elements ranged between 324 and 40,670 mg/Kg fish, while the average values found for the most predominant trace elements in terms of concentration ranged from 2196 to 17,058 µg/Kg fish. On the other hand, the levels of Cd, Sb, U, and Tl were below the LOQ. The rest of the elements showed intermediate levels compared to the aforementioned ones.

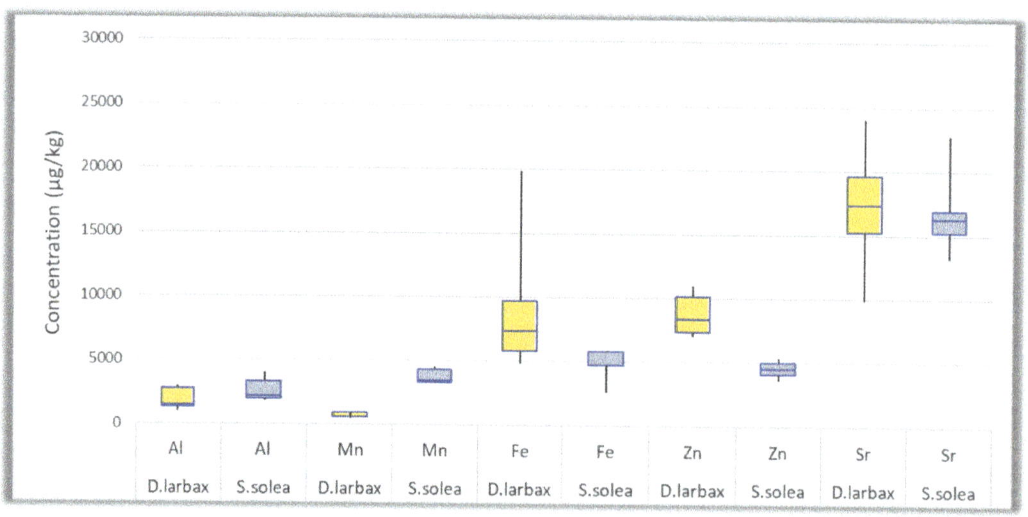

Figure 4. Box plots of the most abundant trace elements (Al, Mn, Fe, Zn, Sr) concentration (µg/Kg) in *D. larbax* (n = 5) and *S. solea* (n = 5) fish samples.

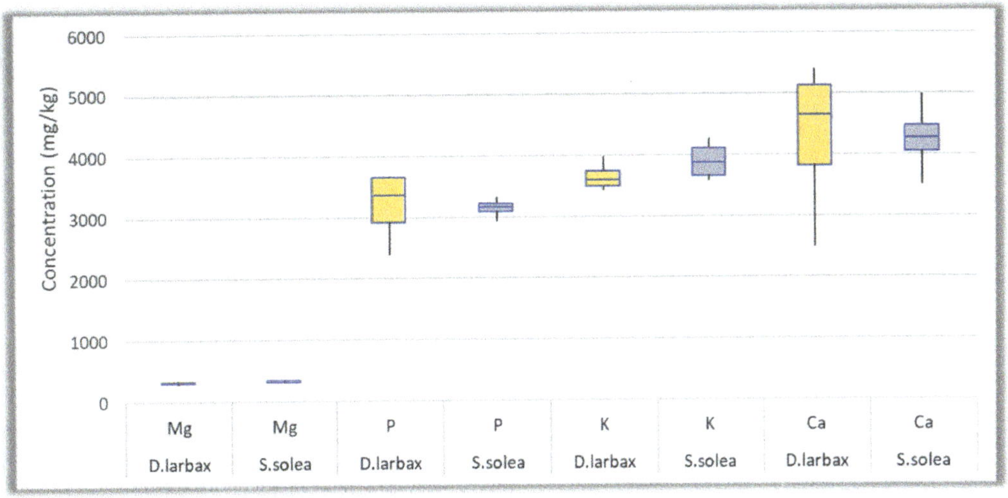

Figure 5. Box plots of macro elements concentration (mg/Kg) in *D. larbax* (n = 5) and *S. solea* (n = 5) fish samples.

The trace elements, characterized by high toxicity (Pb, Hg, Cd, and total As) were all detected in low concentrations or were not detected at all (concentrations ranged between <LOQ to 128 μg/Kg) in both fish species (Figure 6). This was also the case for the elements that can be both essential and toxic, depending on their oxidation state and concentration (Ni, Cr, and Cu). More specifically, regarding the three latter elements, their detected values were below 873 μg/Kg in all the analyzed samples, except for one sample (sample code: 1, *D. larbax*) that was found to be more contaminated for Ni (Figure 6). With regard to the established maximum levels of Pb (300 μg/Kg fish), Hg (500 μg/Kg fish), and Cd (50 μg/Kg fish) in fish [47], no exceedance was observed in the analyzed samples, although the LOQs for Pb, Hg, and Cd were below the MRLs (see Table S6).

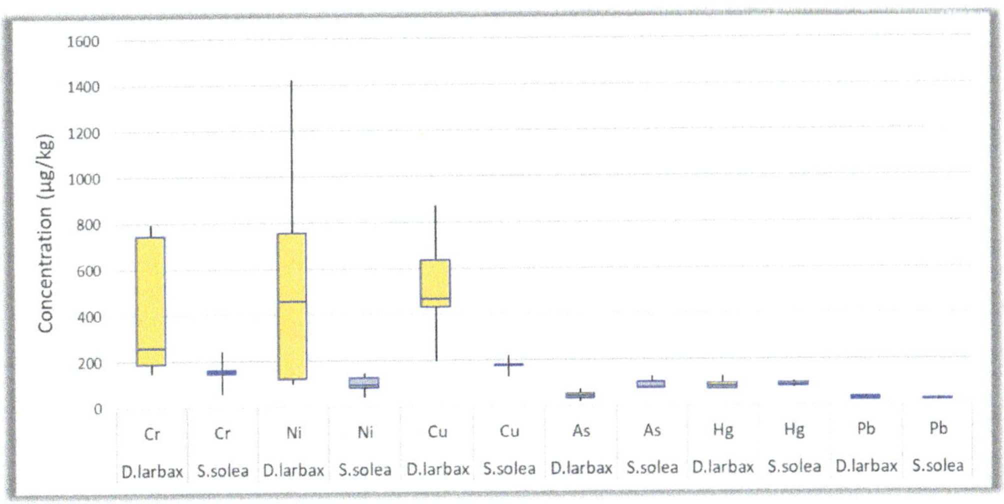

Figure 6. Box plots of toxic elements concentration (μg/Kg) in *D. larbax* (n = 5) and *S. solea* (n = 5) fish samples. Cd concentrations are not included, since they were all below LOQ value.

By comparing the contamination levels between the two fish species through the application of an ANOVA test ($p < 0.05$), it was observed that the detected concentrations were of comparable range for all the analytes, except for Cu ($p = 0.015846$), Mn ($p = 3.23 \times 10^{-6}$), As ($p = 9.85 \times 10^{-3}$, and Zn ($p = 9.92 \times 10^{-4}$). In particular, *S. solea* samples were found to have higher concentrations of Mn, while the detected values of Zn and Cu were elevated in *D. larbax*. More specifically, the average concentrations of Mn in *S. solea* were equal to 3785 μg/Kg fish, while in *D. larbax* were 607 μg/Kg fish. The average concentration of As in *S. solea* fish (92 μg/Kg) was approximately twice as high as the concentration detected in *D. larbax* species (46 μg/Kg fish). The opposite was the case for Zn and Cu (average concentrations: 8780 and 522 μg/Kg fish, versus 4538 and 176 μg/Kg fish in *D. larbax* and *S. soela*, respectively). However, the limited number ($n = 10$) of fish samples did not allow us to draw a solid conclusion, and thus further investigation is needed. *S. solea* samples showed higher concentrations for the elements Al, Mn, Fe, Zn, and Sr (range of mean concentrations between 2.7 and 17 mg/Kg), compared to the rest of the trace analytes (mean concentrations below 1 mg/Kg). In another study examining minerals in the tissues of *S. solea* collected from Thermaikos Gulf, it was found that the highest detected levels of elements were observed in the gills, the liver, and the scales of the samples, while the lowest was in their muscle [20]. Since the current fish samples were homogenized, and their tissues were not separated, a direct comparison with the results presented by Vetsis and co-workers [20], was not possible.

3.5. Human Health Risk Assessment

Health risk assessment associated with fish consumption and the occurrence of toxic elements, pesticides, and pharmaceuticals residues was estimated, and the values of ADD and HQ are presented in Table 1. To calculate the latter, the TDI (μg/kg/d) values used for all chemicals were retrieved in principle from EFSA and the open literature.

Table 1. Calculated ADD and HQ values for all the detected pesticides, pharmaceuticals, and toxic elements, considering the worst cases.

				ADD (μg/Kg/day) *		HQ **	
Elements	**TDI (μg/Kg/day)**	**Concentration (μg/Kg)**		**Adults**	**Children**	**Adults**	**Children**
Pb	2 [48]	38		0.03	0.14	0.02	0.07
As	0.3 [49]	125		0.10	0.47	0.33	1.56
Cr	3 [50]	796		0.64	2.97	0.21	0.99
Ni	13 [51]	1422		1.14	5.31	0.09	0.41
Hg	0.1 [52]	128		0.10	0.48	1.02	4.76
				ADD (μg/Kg/day)		**HQ**	
Compounds	**TDI (μg/Kg/day)**	**Concentration (μg/Kg fish)**	**Concentration (μg/Kg d.w.)**	**Adults**	**Children**	**Adults**	**Children**
Thiabendazole	100 [53]	13.7	68.5	5.48×10^{-2}	2.56×10^{-1}	5.48×10^{-4}	2.56×10^{-3}
Cypermethrin	5 [54]	8.7	43.5	3.48×10^{-2}	1.62×10^{-1}	6.96×10^{-3}	3.25×10^{-2}
Tricyclazole	50 [55]	0.9	4.5	3.60×10^{-3}	1.68×10^{-2}	7.20×10^{-5}	3.36×10^{-4}
Caffeine	2.5 [56]	3.8	19	1.52×10^{-2}	7.09×10^{-2}	6.08×10^{-3}	2.84×10^{-2}
Acetaminophen	300 [56]	1.5	7.5	6.00×10^{-3}	2.80×10^{-2}	2.00×10^{-5}	9.33×10^{-5}

* ADD: Average daily intake. ** HQ: Hazard quotient.

For pesticides and pharmaceuticals, concentrations (μg/Kg fish) were converted to dry weight by multiplying by a factor of 5 [57]. More specifically, the reported HQ values were computed based on the uppermost detected concentrations (worst-case scenario) of the toxic elements and compounds quantified in the analyzed samples. In the same context, the admission of regarding non-edible tissues', exemplified by the gastrointestinal tract (a reservoir of chemicals [58,59]), contribution to the detected concentrations was made (the sampled fishes were not gutted). Taking these into consideration, in almost all cases, calculated HQ values of elements for adults were below the threshold value of 1, indicating that the consumption of fish does not pose any risk to adults. The only exception accounted for one sample for Hg (HQ = 1.02). On the contrary, when children were considered, HQ values for Hg exceeded the threshold level of 1, which signifies a potential threat. For As in two cases out of ten, HQ values for total As in samples 7 and 8 were found to exceed the threshold level (Table S6).

4. Discussion

In view of the above results, differences in concentrations of macro and trace elements and their patterns have been reported in the literature between benthic, demersal, and pelagic fishes [60]. Therefore, the etiology of differences in Zn and Mn levels between the two species can also be attributed to the distinction between *S. solea* (benthic as juvenile and adult) and *D. labrax* (manifests demersal behavior but is more common in shallow waters [61]) and the difference of marine habitat in association to the depth. In a work conducted in the Galapagos marine reserve [60], higher levels of Zn were recorded in pelagic species compared to demersal, which, considering the species collected in this study and their habitat, is in line with the presented findings. In the same context, concentrations of heavy metals were reported to be higher in benthic species compared to pelagic ones [62], possibly due to the proximity of such species to the seabed and the subsequent release and absorption of metals from the sediment (heavy metals usually display higher concentrations at sea bottom compared to the surface water [63]). The latter is in agreement with the presented results for As, in which both the individual and average concentrations in *S. solea* samples were higher than the respective ones in *D. labrax*. For Pb, such a trend was not observed, while for Cd, all values were <LOQ. Nevertheless, based on the bibliography [64], observations should be viewed and deciphered with caution, since the bioavailability of elements is influenced by ecological processes related to specific regions and seasons that can reverse the expected bioaccumulation potential of elements in fishes.

At this point, it is essential to mention that the random and limited sampling (n = 10) of the fishes investigated in this study should not be used to draw generic conclusions for the sampled area and fishes and the potential threat to human health. In the same context, Hg (and methyl mercury, CH_3Hg) is a well-debated topic in the frame of risks and benefits of fish consumption at the forefront of pertinent authorities such as EFSA [65]. As is an element of toxicological concern, with several inorganic forms that are highly soluble in water exhibiting substantial mobility [66], and with numerous sources [67] such as geothermal volcanic activities [68,69], metal mining [70], and other arsenic-generated waste [71]. For Greece in particular, in the islands of the Aegean Sea, there is still active mining and volcanic activity, which, along with the mobility of seawater and capacity to transfer chemicals in large distances, can justify the occurrence and subsequent accumulation of As and other metals in fish tissues [69,72]. In the same context, the topical release of contaminants by ships due to the unintentional release of waste (that contains As and other pollutants) during transportation is also reported, and can be another source of pollution. Nevertheless, the number of analyzed samples is limited, and thus, a more extended study is recommended for a more reliable outcome of risk assessment to be drawn and to better portray the residual prevalence of the sampled area. From the health risk assessment perspective, the inclusion of non-edible tissues in the sample preparation might have impacted the reported concentrations (and findings), especially considering the gastrointestinal tract content known to accumulate chemicals. Nevertheless, this source of uncertainty seems to be "considered" when viewed in the context of a worst-case scenario and the likelihood of slightly elevated concentrations' insertion in risk assessment calculations.

Computational toxicology and in silico calculations can be a valuable tool in the context of standard approaches used in toxicity prediction, mechanistic rationale, and health risk assessment. Perin and colleagues used a similar approach in the assessment of organic compounds detected in surface water [73]. In the present work, and in the context of demonstrating the importance of in silico toxicity evaluation, using the Derek Nexus in silico tool for both acetaminophen and caffeine, an informative overview of the potentially toxic effects of those compounds was achieved, which can be partially contrasted and discussed with the concentration found in fish samples. Among the 62 endpoints investigated via the in silico tool, 54 did not result in any alarming effect at the selected reasoning level. Exact matches for acetaminophen were reported for two endpoints. Acetaminophen has been implicated (as a potential causative agent) in hepatotoxicity effects after long-term use and intentional dosing by ingestion (i.e., >4 g/day, as maximum dose advocated on the package) [74]. Even though they are not directly comparable (administered levels of pharmaceuticals in contrast to traces detected in fish commodity), the herein detected levels for both compounds are far below the concentrations reported; therefore, realistic consumption of such fish will possibly not lead to an uptake of such high concentrations. Overall, the in silico toxicity tool adds another evidence to the outcome corroborated in the HQ calculations.

5. Conclusions

A battery of analytical methods covering more than 280 chemicals exemplified by pesticides, pharmaceuticals, and macro and trace elements was applied to fish samples collected from Thermaikos Gulf, one of the major fishing grounds in Greece. The low prevalence of pesticides and pharmaceuticals was corroborated, with overall concentrations not exceeding 14 μg/Kg fish. The latter signifies that despite the intensive agricultural and industrial activity in the extended coastal area of Thermaikos Gulf, the chemical concentrations still remain low. As a general observation, we underline that the results from a human health risk assessment perspective are of particular importance due to the presence of some heavy metals, considering the potential threat that may pose to children after the long-term consumption of fish. However, in no case these results can be used for actual and reliable risk–benefit analysis of fish consumption for human health due to the very limited number of samples and lack of any systematic monitoring methodology. Future steps include the extended sampling of fish species, the incorporation of CRM negative control fish tissue samples that will strengthen the validation of the analytical methods and decrease analytical uncertainty, further extension of the analytical scope to pesticides and pharmaceuticals' metabolites, the speciation analysis of As and Cr (both inorganic and organic forms), and the untargeted chemical analysis via high-resolution mass spectrometry systems to address the totality of pollutants in the specific marine environment.

Supplementary Materials: The following supporting information can be downloaded at: https://www.mdpi.com/article/10.3390/jox14020041/s1, Table S1: Analytical scope (pesticides), LOQs and recoveries (at the LOQ) obtained in fish; Table S2: LODs, LOQs, and recoveries (including standard deviation, SD) obtained for pharmaceuticals in whole fish; Table S3: MRM conditions for antibiotics and pharmaceuticals in LC-ESI-MS/MS; Table S4: Concentrations of trace elements (mg/kg) in each individual fish sample; Table S5: Concentrations of macro elements (mg/kg) in each individual fish sample; Table S6: Quantification parameters of trace and macro elements; Table S7: Calculated HQ (adult, children) values for the most toxic elements; Figure S1: MRM chromatograms of *o,p*′-DDE traces (in proximity to the LOD) in seabass; Figure S2: MRM chromatograms of caffeine in: (a) blank fish sample, (b) in spiked fish sample at 5 μg/Kg fish, and (c) MRM chromatogram (quantitation) of ibuprofen traces (at LOD) in *D. labrax*.

Author Contributions: Conceptualization, D.K., N.D. and K.M.; methodology, K.M.K. and E.Z.; software, K.M.K., E.Z., Z.S. and F.T.; validation, K.M.K., E.Z. and E.M.-K.; formal analysis, K.M.K. and E.Z.; investigation, K.M.K., E.Z., F.T., Z.S., D.K., N.D. and K.M.; resources, D.K., N.D. and K.M.; data curation, K.M.K., E.Z. and E.M.-K.; writing—original draft preparation, K.M.K. and E.Z.; writing—review and editing, K.M.K., E.Z. and K.M.; visualization, E.Z. and E.M.-K.; supervision, D.K. and K.M.; project administration, D.K., K.M. and K.M.K.; funding acquisition, N.D. and K.M. All authors have read and agreed to the published version of the manuscript.

Funding: The analytical work was covered by internal resources of Benaki Phytopathological Institute.

Institutional Review Board Statement: Not applicable.

Informed Consent Statement: Not applicable.

Data Availability Statement: All data are presented in the manuscript and its supplementary material.

Acknowledgments: The sampling was conducted in the context of FishMedic program, implemented within the framework of the "Support Points" program, co-financed by the John S. Latsis Public Benefit Foundation, the TIMA Charitable Foundation the Hellenic Hope Charitable Organization, the Captain Vassilis & Carmen Constantakopoulos Foundation, A.G Leventis Foundation, the A. C. Laskaridis Charitable Foundation, the Costas M. Lemos Foundation and the Bodosaki Foundation.

Conflicts of Interest: The authors declare no conflicts of interest.

References

1. Nunes, M.; Leston, S. Coastal Pollution: An Overview. In *Life Below Water*; Encyclopedia of the UN Sustainable Development Goals; Springer: Cham, Switzerland, 2020; pp. 1–11.
2. Directive-2008/56/EC. *Directive 2008/56/EC of the European Parliament and of the Council: Establishing a Framework for Community Action in the Field of Marine Environmental Policy (Marine Strategy Framework Directive)*; Official Journal of the European Union: Luxembourg, 2008.
3. Madikizela, L.M.; Ncube, S.; Tutu, H.; Richards, H.; Newman, B., Ndungu, K.; Chimuka, L. Pharmaceuticals and their metabolites in the marine environment: Sources, analytical methods and occurrence. *Trends Environ. Anal.* **2020**, *28*, e00104. [CrossRef]
4. Nödler, K.; Voutsa, D.; Licha, T. Polar organic micropollutants in the coastal environment of different marine systems. *Mar. Pollut. Bull.* **2014**, *85*, 50–59. [CrossRef] [PubMed]
5. Ali, A.M.; Ronning, H.T.; Alarif, W.; Kallenborn, R.; Al-Lihaibi, S.S. Occurrence of pharmaceuticals and personal care products in effluent-dominated Saudi Arabian coastal waters of the Red Sea. *Chemosphere* **2017**, *175*, 505–513. [CrossRef] [PubMed]
6. Shahjahan, M.; Taslima, K.; Rahman, M.S.; Al-Emran, M.; Alam, S.I.; Faggio, C. Effects of heavy metals on fish physiology—A review. *Chemosphere* **2022**, *300*, 134519. [CrossRef] [PubMed]
7. Javed, M.; Usmani, N. An overview of the adverse effects of heavy metal contamination on fish health. *Proc. Natl. Acad. Sci. USA* **2019**, *89*, 389–403. [CrossRef]
8. Rohani, M.F. Pesticides toxicity in fish: Histopathological and hemato-biochemical aspects—A review. *Emerg. Contam.* **2023**, *9*, 100234. [CrossRef]
9. Emmanouil, C.; Smart, D.J.; Hodges, N.J.; Chipman, J.K. Oxidative damage produced by Cr(VI) and repair in mussel (*Mytilus edulis* L.) gill. *Mar. Environ. Res.* **2006**, *62*, S292–S296. [CrossRef] [PubMed]
10. Lee, J.W.; Choi, H.; Hwang, U.K.; Kang, J.C.; Kang, Y.J.; Kim, K.I.; Kim, J.H. Toxic effects of lead exposure on bioaccumulation, oxidative stress, neurotoxicity, and immune responses in fish: A review. *Environ. Toxicol. Pharmacol.* **2019**, *68*, 101–108. [CrossRef]
11. Martínez-Megías, C.; Arenas-Sánchez, A.; Manjarrés-López, D.; Pérez, S.; Soriano, Y.; Picó, Y.; Rico, A. Pharmaceutical and pesticide mixtures in a Mediterranean coastal wetland: Comparison of sampling methods, ecological risks, and removal by a constructed wetland. *Environ. Sci. Pollut. Res.* **2024**, *31*, 14593–14609. [CrossRef]
12. Palma, P.; Fialho, S.; Lima, A.; Catarino, A.; Costa, M.J.; Barbieri, M.V.; Monllor-Alcaraz, L.S.; Postigo, C.; de Alda, M.L. Occurrence and risk assessment of pesticides in a Mediterranean Basin with strong agricultural pressure (Guadiana Basin: Southern of Portugal). *Sci. Total Environ.* **2021**, *794*, 148703. [CrossRef]
13. Ardila, P.A.R.; Alvarez-Alonso, R.; Arcega-Cabrera, F.; Valsero, J.J.D.; García, R.M.; Lamas-Cosío, E.; Oceguera-Vargas, I.; DelValls, A. Assessment and review of heavy metals pollution in sediments of the Mediterranean Sea. *Appl. Sci.* **2024**, *14*, 1435. [CrossRef]
14. Dimarchopoulou, D.; Tsagarakis, K.; Sylaios, G.; Tsikliras, A.C. Ecosystem trophic structure and fishing effort simulations of a major fishing ground in the northeastern Mediterranean Sea (Thermaikos Gulf). *Estuar. Coast. Shelf Sci.* **2022**, *264*, 107667. [CrossRef]
15. iSea-Thermaikos. Available online: https://isea.com.gr/thermaikos-gulf-mpa/?lang=en (accessed on 2 September 2023).

16. Markantonatou, V.; Giakoumi, S.; Koukourouvli, N.; Maina, I.; Gonzalez-Mirelis, G.; Sini, M.; Maistrelis, K.; Stithou, M.; Gadolou, E.; Petza, D.; et al. Marine spatial plans focusing on biodiversity conservation: The case of the Aegean Sea. *Aquat. Conserv.* **2021**, *31*, 2278–2292. [CrossRef]
17. Dimarchopoulou, D.; Dogrammatzi, A.; Karachle, P.K.; Tsikliras, A.C. Spatial fishing restrictions benefit demersal stocks in the northeastern Mediterranean Sea. *Sci Rep.* **2018**, *8*, 5967. [CrossRef]
18. Stergiou, K.; Moutopoulos, D.K.; Tsikliras, A. Spatial and temporal variability in Hellenic marine fisheries landings: State of Hellenic fisheries. In *State of Hellenic Fisheries*; Papaconstantinou, C., Zenetos, A., Vassilopoulou, V., Tserpes, G., Eds.; Hellenic Centre for Marine Research: Athens, Greece, 2007; pp. 141–150.
19. Sylaios, G.K.; Koutroumanidis, T.; Tsikliras, A.C. Ranking and classification of fishing areas using fuzzy models and techniques. *Fisheries Manag. Ecol.* **2010**, *17*, 240–253. [CrossRef]
20. Vetsis, E.; Kalantzi, I.; Pergantis, S.A.; Kokokiris, L.; Karakassis, I. Metals in tissues of marine fish from the Thermaikos Gulf, Eastern Mediterranean Sea: Detection of changes with trophic level. *Mar. Pollut. Bull.* **2021**, *173*, 113024. [CrossRef] [PubMed]
21. Catsiki, V.A.; Florou, H. Study on the behavior of the heavy metals Cu, Cr, Ni, Zn, Fe, Mn and Cs in an estuarine ecosystem using as a bioindicator species: The case of Thermaikos gulf, Greece. *J. Environ. Radioactiv.* **2006**, *86*, 31–44. [CrossRef]
22. Pell, A.; Kokkinis, G.; Malea, P.; Pergantis, S.A.; Rubio, R.; Lopez-Sanchez, J.F. LC-ICP-MS analysis of arsenic compounds in dominant seaweeds from the Thermaikos Gulf (Northern Aegean Sea, Greece). *Chemosphere* **2013**, *93*, 2187–2194. [CrossRef]
23. Christophoridis, C.; Dedepsidis, D.; Fytianos, K. Occurrence and distribution of selected heavy metals in the surface sediments of Thermaikos Gulf, N. Greece. Assessment using pollution indicators. *J. Hazard. Mater.* **2009**, *168*, 1082–1091. [CrossRef]
24. Christophoridis, C.; Bourliva, A.; Evgenakis, E.; Papadopoulou, L.; Fytianos, K. Effects of anthropogenic activities on the levels of heavy metals in marine surface sediments of the Thessaloniki Bay, Northern Greece: Spatial distribution, sources and contamination assessment. *Microchem. J.* **2019**, *149*, 104001. [CrossRef]
25. Malea, P.; Rijstenbil, J.W.; Haritonidis, S. Effects of cadmium, zinc and nitrogen status on non-protein thiols in the macroalgae spp. from the Scheldt Estuary (SW Netherlands, Belgium) and Thermaikos Gulf (N Aegean Sea, Greece). *Mar. Environ. Res.* **2006**, *62*, 45–60. [CrossRef] [PubMed]
26. Haritonidis, S.; Malea, P. Bioaccumulation of metals by the green alga from Thermaikos Gulf, Greece. *Environ. Pollut.* **1999**, *104*, 365–372. [CrossRef]
27. Albanis, T.A.; Danis, T.G.; Kourgia, M.K. Transportation of pesticides in estuaries of the Axios, Loudias and Aliakmon rivers (Thermaikos Gulf), Greece. *Sci. Total Environ.* **1994**, *156*, 11–22. [CrossRef]
28. FishBase; Froese, R.; Pauly, D. (Eds.) World Wide Web Electronic Publication. Version 02/2024. 2024. Available online: www.fishbase.org (accessed on 15 February 2024).
29. Fish-Sampling. Available online: https://environment.des.qld.gov.au/__data/assets/pdf_file/0031/90787/biological-assessment-fish-collection-and-the-dissection-for-the-purpose-of-chemical-analysis-of-tissues.pdf (accessed on 2 April 2021).
30. Kasiotis, K.M.; Anagnostopoulos, C.; Anastasiadou, P.; Machera, K. Pesticide residues in honeybees, honey and bee pollen by LC-MS/MS screening: Reported death incidents in honeybees. *Sci. Total Environ.* **2014**, *485–486*, 633–642. [CrossRef]
31. Kasiotis, K.M.; Tzouganaki, Z.D.; Machera, K. Chromatographic determination of monoterpenes and other acaricides in honeybees: Prevalence and possible synergies. *Sci. Total Environ.* **2018**, *625*, 96–105. [CrossRef]
32. Kasiotis, K.M.; Zafeiraki, E.; Kapaxidi, E.; Manea-Karga, E.; Antonatos, S.; Anastasiadou, P.; Milonas, P.; Machera, K. Pesticides residues and metabolites in honeybees: A Greek overview exploring Varroa and Nosema potential synergies. *Sci. Total Environ.* **2021**, *769*, 145213. [CrossRef] [PubMed]
33. Zafeiraki, E.; Kasiotis, K.M.; Nisianakis, P.; Manea-Karga, E.; Machera, K. Occurrence and human health risk assessment of mineral elements and pesticides residues in bee pollen. *Food Chem. Toxicol.* **2022**, *161*, 112826. [CrossRef] [PubMed]
34. Shaaban, H.; Mostafa, A. Simultaneous determination of antibiotics residues in edible fish muscle using eco-friendly SPE-UPLC-MS/MS: Occurrence, human dietary exposure and health risk assessment for consumer safety. *Toxicol. Rep.* **2023**, *10*, 1–10. [CrossRef]
35. Vitale, D.; Pico, Y.; Alvarez-Ruiz, R. Determination of organic pollutants in *Anguilla anguilla* by liquid chromatography coupled with tandem mass spectrometry (LC-MS/MS). *MethodsX* **2021**, *8*, 101342. [CrossRef]
36. Drabova, L.; Dvorakova, D.; Urbancova, K.; Gramblicka, T.; Hajslova, J.; Pulkrabova, J. Critical assessment of clean-up techniques employed in simultaneous analysis of persistent organic pollutants and polycyclic aromatic hydrocarbons in fatty samples. *Toxics* **2022**, *10*, 12. [CrossRef]
37. Zafeiraki, E.; Sabo, R.; Kasiotis, K.M.; Machera, K.; Sabova, L.; Majchrak, T. Adult honeybees and beeswax as indicators of trace elements pollution in a vulnerable environment: Distribution among different apicultural compartments. *Molecules* **2022**, *27*, 6629. [CrossRef] [PubMed]
38. SANTE/11312/2021. Guidance Document on Analytical Quality Control and Method Validation Procedures for Pesticides Residues Analysis in Food and Feed. Available online: https://www.eurl-pesticides.eu/userfiles/file/EurlALL/SANTE_11312_2021.pdf (accessed on 5 June 2021).
39. *US-EPA-2000*; Risk Based Concentration Table; United States Environmental Protection Agency: Philadelphia, PA, USA. Available online: https://archive.epa.gov/region9/superfund/web/html/index-23.html (accessed on 1 September 2023).
40. Fish-Consumption. Available online: https://oceans-and-fisheries.ec.europa.eu/facts-and-figures/facts-and-figures-common-fisheries-policy/consumption_en (accessed on 6 January 2024).

41. Barnhoorn, I.; van Dyk, C. The first report of selected herbicides and fungicides in water and fish from a highly utilized and polluted freshwater urban impoundment. *Environ. Sci. Pollut. Res. Int.* **2020**, *27*, 33393–33398. [CrossRef] [PubMed]
42. Ccanccapa, A.; Masia, A.; Andreu, V.; Pico, Y. Spatio-temporal patterns of pesticide residues in the Turia and Jucar Rivers (Spain). *Sci. Total Environ.* **2016**, *540*, 200–210. [CrossRef] [PubMed]
43. EU-MRLs. Available online: https://ec.europa.eu/food/plant/pesticides/eu-pesticides-database/start/screen/mrls (accessed on 2 October 2023).
44. Codex-Alimentarius. Available online: https://www.fao.org/fao-who-codexalimentarius/codex-texts/dbs/pestres/pesticides/en/ (accessed on 2 October 2023).
45. US-EPA-MRLs. Available online: https://www.ecfr.gov/current/title-40/chapter-I/subchapter-E/part-180 (accessed on 2 October 2023).
46. Australian-MRLs. Available online: https://www.legislation.gov.au/F2023L01350/latest/versions (accessed on 2 October 2023).
47. EU-REG-2023/915. Commission Regulation (EU) 2023/915 of 25 April 2023 on Maximum Levels for Certain Contaminants in Food and Repealing Regulation (EC) No 1881/2006. Available online: https://eur-lex.europa.eu/legal-content/EN/TXT/PDF/?uri=CELEX:32023R0915 (accessed on 2 February 2024).
48. Lead-EFSA. Scientific Opinion on Lead in Food. *Efsa J.* **2010**, *8*, 1570.
49. Arsenic-EFSA. Scientific Opinion on Arsenic in Food. *Efsa J.* **2009**, *7*, 1351. [CrossRef]
50. Chromium-EFSA. Scientific Opinion on the risks to public health related to the presence of chromium in food and drinking water: EFSA Panel on Contaminants in the Food Chain (CONTAM). *Efsa J.* **2014**, *12*, 3595.
51. Schrenk, D.; Bignami, M.; Bodin, L.; Chipman, J.K.; del Mazo, J.; Grasl-Kraupp, B.; Hogstrand, C.; Hoogenboom, L.; Leblanc, J.C.; Nebbia, C.S.; et al. Update of the risk assessment of nickel in food and drinking water. *Efsa J.* **2020**, *18*, e06268. [CrossRef] [PubMed]
52. Mercury-EFSA. Scientific Opinion on the risk for public health related to the presence of mercury and methylmercury in food. *EFSA J.* **2012**, *10*, 2985.
53. Bellisai, G.; Bernasconi, G.; Brancato, A.; Cabrera, L.C.; Castellan, I.; Ferreira, L.; Giner, G.; Greco, L.; Jarrah, S.; Leuschner, R.; et al. Evaluation of confirmatory data following the Article 12 MRL review for thiabendazole. *Efsa J.* **2022**, *20*, e07539. [CrossRef]
54. Arena, M.; Auteri, D.; Barmaz, S.; Brancato, A.; Brocca, D.; Bura, L.; Cabrera, L.C.; Chiusolo, A.; Civitella, C.; Marques, D.C.; et al. Peer review of the pesticide risk assessment of the active substance cypermethrin. *Efsa J.* **2018**, *16*, e05402. [CrossRef]
55. Bellisai, G.; Bernasconi, G.; Brancato, A.; Cabrera, L.C.; Castellan, I.; Del Aguila, M.; Ferreira, L.; Santonja, G.G.; Greco, L.; Jarrah, S.; et al. Setting of import tolerance for tricyclazole in rice. *Efsa J.* **2023**, *21*, e07757. [CrossRef]
56. Sorell, T.L. Approaches to the Development of Human Health Toxicity Values for Active Pharmaceutical Ingredients in the Environment. *AAPS J.* **2016**, *18*, 92–101. [CrossRef]
57. Weight-Conversion-Factor. Available online: http://www.cresp.org/Amchitka/Final_WW_DW_3_13_06.pdf (accessed on 2 May 2023).
58. Fatema, K.; Islam, M.J.; Sarker, M.A.I.; Elahi, K.S.; Alam, M.J.; Hasan, S.J.; Rashid, H. Occurrence of microplastics in fish gastrointestinal tracts belongs to different feeding habits from the Bangladesh coast of the Bay of Bengal. *Environ. Sci. Pollut. Res. Int.* **2024**, *31*, 24329–24343. [CrossRef] [PubMed]
59. SANTE/10254/2021. Nature of Pesticides Residues in Fish. Available online: https://food.ec.europa.eu/document/download/3cb924c2-a651-45d0-89d7-7ce8de6b6dd1_en?filename=pesticides_mrl_guidelines_app-j-10254-2021_en.pdf (accessed on 5 May 2024).
60. Hossain, M.B.; Tanjin, F.; Rahman, M.S.; Yu, J.; Akhter, S.; Abu Noman, M.; Sun, J. Metals bioaccumulation in 15 commonly consumed fishes from the lower Meghna River and adjacent areas of Bangladesh and associated human health hazards. *Toxics* **2022**, *10*, 139. [CrossRef]
61. Lloris, D. A world overview of species of interest to fisheries. Chapter: Dicentrarchus labrax: 3p. FIGIS Species Fact Sheets. In *Species Identification and Data Programme-SIDP*; FAO-FIGIS, 2002; Available online: https://www.fao.org/fishery-divisional-structure/en (accessed on 1 May 2024).
62. Amankwaa, G.; Lu, Y.; Liu, T.; Wang, N.; Luan, Y.; Cao, Y.; Huang, W.; Ni, X.; Gyimah, E. Heavy metals concentration profile of an aquatic environment and health implications of human exposure to fish and prawn species from an urban river (Densu). *Iran J. Fish Sci.* **2021**, *20*, 529–546. [CrossRef]
63. Ismukhanova, L.; Choduraev, T.; Opp, C.; Madibekov, A. Accumulation of heavy metals in bottom sediment and their migration in the water ecosystem of Kapshagay reservoir in Kazakhstan. *Appl. Sci.* **2022**, *12*, 11474. [CrossRef]
64. Rejomon, G.; Nair, M.; Joseph, T. Trace metal dynamics in fishes from the southwest coast of India. *Environ. Monit. Assess.* **2010**, *167*, 243–255. [CrossRef]
65. EFSA-Scientific-Committee. SCIENTIFIC OPINION: Statement on the benefits of fish/seafood consumption compared to the risks of methylmercury in fish/seafood. *EFSA J.* **2015**, *13*, 3982. [CrossRef]
66. Neff, J.M. Arsenic in the Ocean. In *Bioaccumulation in Marine Organisms: Effect of Contaminants from Oil Well Produced Water*; Elsevier Science: Amsterdam, The Netherlands, 2002; pp. 57–78.
67. Smedley, P.L.; Kinniburgh, D.G. A review of the source, behaviour and distribution of arsenic in natural waters. *Appl. Geochem.* **2002**, *17*, 517–568. [CrossRef]

68. Bundschuh, J.; Maity, J.P. Geothermal arsenic: Occurrence, mobility and environmental implications. *Renew. Sust. Energ Rev.* **2015**, *42*, 1214–1222. [CrossRef]
69. Golfinopoulos, S.K.; Varnavas, S.P.; Alexakis, D.E. The status of arsenic pollution in the Greek and Cyprus environment: An overview. *Water* **2021**, *13*, 224. [CrossRef]
70. Martínez-López, S.; Martínez-Sánchez, M.J.; Gómez-Martínez, M.D.; Pérez-Sirvent, C. Arsenic zoning in a coastal area of the Mediterranean Sea as a base for management and recovery of areas contaminated by old mining activities. *Appl. Clay Sci.* **2020**, *199*, 105881. [CrossRef]
71. Wang, N.X.; Ye, Z.J.; Huang, L.P.; Zhang, C.S.; Guo, Y.X.; Zhang, W. Arsenic occurrence and cycling in the aquatic environment: A comparison between freshwater and seawater. *Water* **2023**, *15*, 147. [CrossRef]
72. Pappa, F.K.; Tsabaris, C.; Ioannidou, A.; Patiris, D.L.; Kaberi, H.; Pashalidis, I.; Eleftheriou, G.; Androulakaki, E.G.; Vlastou, R. Radioactivity and metal concentrations in marine sediments associated with mining activities in Ierissos Gulf, North Aegean Sea, Greece. *Appl. Radiat. Isot.* **2016**, *116*, 22–33. [CrossRef] [PubMed]
73. Perin, M.; Dallegrave, A.; da Costa, J.S.; Streit, L.; Gomes, A.D.; Pizzolato, T.M. Identification of the organic compounds in surface water: Suspect screening using liquid chromatography high-resolution mass spectrometry and in silico toxicity evaluation. *Int. J. Mass Spectrom.* **2023**, *484*, 116982. [CrossRef]
74. Larson, A.M.; Polson, J.; Fontana, R.J.; Davern, T.J.; Lalani, E.; Hynan, L.S.; Reisch, J.S.; Schiodt, F.V.; Ostapowicz, G.; Shakil, A.O.; et al. Acetaminophen-induced acute liver failure: Results of a United States multicenter, prospective study. *Hepatology* **2005**, *42*, 1364–1372. [CrossRef]

Journal of *Xenobiotics*

MDPI

Article

Potential Toxicity of Nine Rare Earth Elements (REEs) on Marine Copepod *Tigriopus fulvus*

Francesca Biandolino [1], Ermelinda Prato [1,*], Asia Grattagliano [2], Giovanni Libralato [3], Marco Trifuoggi [4] and Isabella Parlapiano [1]

1 National Research Council, Water Research Institute (IRSA-CNR), Via Roma, 3, 74123 Taranto, Italy; francesca.biandolino@irsa.cnr.it (F.B.); isabella.parlapiano@irsa.cnr.it (I.P.)
2 Department of Chemical Sciences and Technologies, University of Rome "Tor Vergata", Via della Ricerca Scientifica, 1, 00133 Roma, Italy; asia.grattagliano@uniroma2.it
3 Department of Biology, University of Naples Federico II, Via Cinthia 21, 80126 Naples, Italy; giovanni.libralato@unina.it
4 Department of Chemical Sciences, University of Naples Federico II, Via Cinthia 21, 80126 Naples, Italy; marco.trifuoggi@unina.it
* Correspondence: linda.prato@irsa.cnr.it; Tel.: +39-346-1099339

Abstract: The present study focused, for the first time, on the adverse effects of nine REEs on the marine copepod *Tigriopus fulvus*. For this purpose, copepod mortality, immobilization, and naupliar development were assessed. Overall, the results demonstrated that all REEs tested exerted significant adverse effects on *T. fulvus*, with LC50 values ranging from 0.56 to 1.99 mg/L. Concentration-dependent increases in mortality and immobilization for all tested REEs were observed. Following exposure of nauplii to REEs, a significant slowing of nauplii development was shown with all REEs tested. The results obtained clearly highlight the potential toxicity of REEs, and, in particular, of Lanthanum, which could have consequences on the survival and development of *T. fulvus*, affecting the copepod population.

Keywords: rare earth elements (REEs); potential impact; marine copepod; acute mortality tests

1. Introduction

Technological progress in recent decades have led to an increase in demand for rare earth elements (REEs). This group of seventeen metals, including Scandium, Yttrium, and lanthanides, has many extraordinary physico-chemical, catalytic, electrical, magnetic, and optical properties, making them essential in fields ranging from smart and green technologies to raw materials for industrial applications [1,2]. In modern technologies such as mobile phones, televisions, LED light bulbs, and wind turbines, hybrid automobiles, medical devices, hard drives, and defence technologies, they play a key role [3].

Despite their name, these elements are not 'rare' [4], but most of the reserves of these elements are not economically exploitable and their separation is difficult due to their similar chemical properties [5]. China currently holds a dominant position in the global production of REEs (70%), as it has the largest reserves of REEs, followed by the US (14.3%) and Australia (6%) [6].

REEs occur together in natural systems, as minor or major constituents, rather than as individual native metals, and sites in mining-impacted areas have been found to contain REEs at concentrations up to 100 times normal background levels. Therefore, the risks associated with the extraction and production of REEs concern not only the products but also the REE-containing wastes generated during production. These wastes increase the concentration of REEs in the surface environment and have the potential to alter the natural physico-chemical conditions of the soil and water (e.g., pH and redox states). Acidity can lead to the solubilization of otherwise stable REE fractions, thereby increasing their bioavailability and migration into the environment. As a result, high concentrations of REEs have been detected in soils, vegetables and other crops, and even in human bodies, near extraction and processing sites [7–11].

The increased use of REEs combined with their inadequate removal by wastewater treatment plants increases the risk of these substances entering aquatic environments [12,13], with consequent potential hazard to aquatic organisms [14,15]. Gadolinium (Gd) concentrations of up to 1100 mg/L have been found in river discharges from wastewater treatment plants [16], while concentrations of up to 200 ng/L have been found in unpolluted rivers [17,18]. Lanthanum (La) has been reported at concentrations from 0.029 to 12.85 μg/L in surface waters and up to 40.37 μg/L in alluvial aquifers affected by acid mine drainage [19–21]. The concentrations of La and cerium (Ce) in wastewater range from 7.7 to 80.4 μg/L and from 19.4 to 161 μg/L, respectively [22]. In river water, La and Ce concentrations are lower than in wastewater, ranging from 19.7 to 74 ng/L and from 9.67 to 212 ng/L, respectively [22], but on a local basis, they can range from 80 to 200 μg/L, respectively [23].

The understanding of REEs' mechanisms of action in biological systems is not yet fully clear [24]. Recent studies have focused on the adverse effects of REEs on aquatic organisms such as fish and molluscs, demonstrating that REEs can be absorbed and accumulated in organic tissues [25,26]. This may lead to an increase in reactive oxygen species (ROS), inhibition of the antioxidant system and damage to the DNA of exposed aquatic organisms [27,28]. Most extant research has investigated the toxicity of few REEs (e.g., Gd, La, Ce) on freshwater organisms [24,29–31]. So far, data on the REEs' toxicity effects on marine organisms are still limited, mostly derived from acute and chronic toxicity testing on bivalve molluscs, sea urchins, and unicellular algae [32–37]. Given the increasing use of REEs and their presence in the environment, especially in rivers and coastal areas, there is a need to monitor their potential effects on living biota, including bioassays on model organisms.

The harpacticoid copepod *Tigriopus fulvus* is a commonly used model organism for acute toxicity testing due to its ease of breeding in the laboratory and its year-round availability. It has a high sensitivity to a wide range of toxicants and the results of the related tests are reproducible [38].

The aim of this work was to give insight into the potential toxicity of nine REEs: scandium (Sc), yttrium (Y), lanthanum, (La), cerium (Ce), neodymium (Nd), gadolinium (Gd), dysprosium (Dy), holmium (Ho), and erbium (Er) on the marine copepod *T. fulvus*. Comparisons between the toxicity of the nine elements and their effects on mortality, immobility, and naupliar development were assessed.

2. Materials and Methods

2.1. Preparation of Test Medium, Rare Earth Elements (REE)

Test solutions were freshly prepared before each experiment by diluting commercial analytical grade standard solutions (1 g L^{-1}) of Nd, Ho, Er, Dy, Sc, Y, Ce, La, and Gd (purity > 99%, Agilent, Italy) to the final exposure concentrations by using filtered natural seawater (FNSW, 0.22 μm, 38 PSU). Since the initial acid pH of the standard solutions was outside the tolerance range of the test organism, all pH values of the test solutions were measured with a pH metre (Mettler Toledo Five Easy, Milan, Italy) and adjusted to 8.0 ± 0.5 using NaOH prior to the assays.

For each REE, seven different test solutions in a geometric nominal concentration series with a factor of two (0.10, 0.20, 0.40, 0.80, 1.60, 3.20, and 6.40 mg REE/L) were tested.

Actual concentrations were determined by Inductively Coupled plasma mass spectrometry (ICP-MS, Aurora Bruker M90, Bremen, Germany) according to an established protocol [35]. Triplicate analyses were carried out.

2.2. Experimental Design and Conditions

Test organisms were obtained from a massive culture maintained in filtered natural seawater (FNSW, 0.45 μm) at a salinity of 38 PSU in a temperature-controlled room at 20 ± 2 °C, with cool light (500–1200 lux, Ostram Fluora, L58W/77; Milano, Italy) and a photoperiod cycle of 16 h light and 8 h dark. Tetramarin® (fish food) was mixed with the microalgae *Tetraselmis suecica* and *Isochrysis galbana* to provide a weekly ad libitum diet for the copepods.

2.3. Test Procedures

All tests were carried out on newborns (nauplii ≤ 24 h-old) of *T. fulvus* following protocols reported by Faraponova et al. [38] and UNICHIM [39], in 12-well culture plates; each well contained 10 nauplii in 3 mL of test solution and a negative control (with only FNSW). Three replicates for each concentration and the negative control were performed. Plates were incubated for 96 h under the same conditions described above for culture maintenance without feeding.

The organism's sensitivity and test validity were confirmed by conducting acute tests with the reference compound copper sulphate ($CuSO_4 \times 5\ H_2O$) (positive control).

During the entire duration of the test, the pH (8.0 ± 1.0), salinity (38.0 ± 1.0), and oxygen levels (>80% of saturation) of the solutions were checked.

The measured endpoints were mortality, immobility, and number of naupliar moults.

At the end of the tests, the number of dead and immobile nauplii was counted under a stereomicroscope. When nauplii failed to move any appendages after gentle stimulation for at least 20 s of observation, they were considered dead. The number of immobile nauplii was determined by summing the dead nauplii to those still living, not shifting their barycenter, but moving their appendages.

In addition, the effect of REE on the moulting process of *T. fulvus* was evaluated. The number of exuviae (moults) released during the 96 h test was counted. Crystal violet solution (Sigma-Aldrich, Cas 548-62-9) for microscopy was diluted 1:10 with distilled water. For each chamber test, a drop was added with a Pasteur pipette, to make the exuviae more visible under the stereomicroscope. All tests were replicated three times.

2.4. Data Analyses

All results are expressed as the mean ± standard deviation (SD) (n = 3). The median lethal concentration (LC50) and LC20 and, for immobility, the median effect concentration EC50 and EC20 with relative 95% confidence intervals (C.I.), were calculated using the Litchfield–Wilcoxon method. A smooth function was used to fit concentration-response curves with the GraphPad Prism 8.4.2 programme. The data were examined for normality and homogeneity of variances by Kolmogorov–Smirnov and Bartlett's tests. When both assumptions were satisfied, differences between treatments were assessed via one-way analysis of variance (ANOVA). The post hoc Tukey's test was used to account for differences within groups, setting the statistical significance at $p < 0.05$. All statistical analyses were conducted with the package software Past3 (version 1.0; Hammer, Norway).

3. Results and Discussion

3.1. Chemical Data

Table 1 shows the nominal and measured concentrations of the test solutions determined with the ICP-MS. The ratios of measured to nominal concentrations fell within a very narrow range from 0.64 to 1.22, highlighting that the nominal concentration values were always quite close to the measured ones. Therefore, nominal concentration values were considered reliable.

Table 1. Limit of Detection (LOD) and Limit of Quantitation (LOQ) values are expressed in µg/L. Nominal and measured REE concentrations mg/L (mean ± s.d.; n = 3).

		Nominal Concentrations (mg/L)						
		0.1	**0.2**	**0.4**	**0.8**	**1.6**	**3.2**	**6.4**
	LOD–LOQ (µg/L)	Measured concentrations (mg/L)						
Sc	0.066–0.220	0.08 ± 0.00	0.19 ± 0.01	0.36 ± 0.02	0.70 ± 0.02	1.20 ± 0.04	2.86 ± 0.06	5.57 ± 0.20
Y	0.003–0.010	0.09 ± 0.00	0.22 ± 0.02	0.39 ± 0.05	0.71 ± 0.02	1.45 ± 0.05	2.86 ± 0.03	5.89 ± 0.15
La	0.003–0.009	0.11 ± 0.01	0.18 ± 0.001	0.29 ± 0.002	0.94 ± 0.01	1.02 ± 0.00	3.43 ± 0.01	5.84 ± 0.09
Ce	0.003–0.011	0.09 ± 0.00	0.19 ± 0.01	0.37 ± 0.01	0.77 ± 0.03	1.48 ± 0.02	2.97 ± 0.05	5.60 ± 0.10
Nd	0.004–0.013	0.11 ± 0.00	0.17 ± 0.00	0.26 ± 0.01	0.61 ± 0.02	1.34 ± 0.01	2.70 ± 0.04	5.04 ± 0.10
Gd	0.003–0.009	0.08 ± 0.01	0.21 ± 0.03	0.38 ± 0.02	0.75 ± 0.04	1.60 ± 0.05	3.01 ± 0.06	5.79 ± 0.20
Dy	0.002–0.005	0.1 ± 0.01	0.21 ± 0.01	0.47 ± 0.03	0.93 ± 0.05	1.77 ± 0.10	3.72 ± 0.20	5.99 ± 0.20
Ho	0.002–0.006	0.12 ± 0.01	0.23 ± 0.01	0.42 ± 0.02	1.14 ± 0.05	1.12 ± 0.00	2.86 ± 0.08	6.40 ± 0.38
Er	0.001–0.005	0.12 ± 0.01	0.18 ± 0.01	0.33 ± 0.02	0.92 ± 0.01	1.10 ± 0.05	2.65 ± 0.01	5.73 ± 0.05

3.2. Acute Toxicity Test

The mean percentage of survival in the negative controls was >90% in each experiment, while the LC50 of the positive control was equal to 0.13 (0.07–0.18) mg/L of Cu^{2+}, within the range of the laboratory control chart for this species, and in compliance with the acceptance criteria [40].

In general, the results from this study demonstrated that all tested REEs exerted significant adverse effects on *T. fulvus*. La and Gd exhibited a greater toxicity to *T. fulvus*, with LC50s of 0.56 and 0.92 mg/L, respectively ($p < 0.05$, Table 2), followed by Nd and Ho, which showed similar toxicity. Er was the least toxic at 1.99 mg/L ($p < 0.05$, Table 2). According to LC50s obtained, REEs were ranked in the decreasing order of toxicity as follows: La > Gd > Nd > Ho > Y > Sc > Ce > Dy > Er; therefore, REEs did not show a predictable pattern consistent with their atomic number. Most data on REEs' toxicity to aquatic organisms are derived by acute toxicity tests, although their relative toxicity is often contradictory; therefore, it is unclear whether there is a common pattern of biological effect among all REEs. Some authors reported similar toxicity among REEs [30,41–43], likely due to similar physico-chemical properties influencing chemical behaviour and toxicity mechanisms. Conversely, other studies have shown that each REE has different effects, with different ecotoxicological risks among them. This makes it hard to identify their universal mode of toxicity [24,29,44,45].

Table 2. LC20 and LC50 (mg/L) with 95% C.I., of *T. fulvus* exposed to REEs. Different letters within each column indicate significant differences ($p < 0.05$).

REEs	LC50	C.I. (95%)	LC20	C.I. (95%)
Scandium	1.33 ^{bc}	1.85–0.95	0.44 ^{c}	0.50–0.39
Yttrium	1.17 ^{ab}	1.46–0.93	0.56 ^{d}	0.61–0.50
Lanthanum	0.56 ^{a}	0.74–0.43	0.23 ^{a}	0.25–0.21
Cerium	1.53 ^{bc}	2.08–1.13	0.63 ^{de}	0.71–0.56
Neodynium	1.12 ^{ab}	1.54–0.82	0.39 ^{bc}	0.44–0.35
Gadolinum	0.92 ^{ab}	1.26–0.67	0.32 ^{ab}	0.36–0.28
Dysprosium	1.56 ^{bc}	2.15–1.13	0.62 ^{de}	0.69–0.56
Holmium	1.15 ^{ab}	1.67–0.79	0.33 ^{b}	0.38–0.29
Erbium	1.99 ^{c}	2.86–1.38	0.70 ^{e}	0.80–0.61

Siciliano et al. [46] reported, for *Phaeodactylum tricornutum* exposed to REE values of LC50 from 0.98 to 13.21 mg/L, that the elements's toxicity was ranked as Gd > Ce > Er > La > Eu > Nd > Dy > Sm. Pagano et al. [34], in a study on the effects of several REEs on *Paracentrotus lividus* embryos, reported the following ranking of toxicity: Gd > Y > La >

Nd > Eu > Ce > Sm. On the other hand, no significant differences in the sensitivity of the diatom *Skeletonema costatum* between REEs were observed by Tai et al. [37].

Markich et al. [47] determined the chronic REEs toxicity based on 30 coastal marine organisms, reporting the following ranking: Y~Gd > Lu~Nd~Dy~Ce > La~Pr. As can be seen from the literature, Gd always seems to be one of the most toxic. This is a worrying fact, considering that Gd is the most used worldwide due to the widespread use of Gd-based contrast agents (GBCAs) in medical imaging, and their poor removal efficiency (<10%) by wastewater treatment plants. Indeed, Gd is the most prevalent REE contaminant in coastal marine waters.

The major cause of REE toxicity is the disturbance of Ca metabolism and homeostasis, damaging cells and reducing development, growth, reproduction and survival. This explains why the organisms, which require Ca to build shells, skeletons, or tests, are particularly sensitive to REEs, as reported by Markich et al. [47] for the sea urchin *Heliocidaris tuberculata*.

As in the current study, Borgmann et al. [48] reported for *Hyalella azteca* after one-week exposures to REEs LC50 values between 0.51 and 1.67 mg/L, although in the following decreasing toxicity order: Nd > Y > Gd > Ce > Ho > Dy > Er > La.

Blinova et al. [30] for *Thamnocephalus platyurus* reported LC50 values of 34.6, 33.0, and 31.80 mg/L for La, Ce, and Nd, respectively, while Manusadžianas et al. [49], in a study on 11 lanthanides and Y, found 24 h LC50 values ranging from 3.22 mg/L for Y to 45.2 mg/L for La. Blaise et al. [29], in a study on *Hydra attenuata* exposed to 11 REEs, reported LC50 values in the range 0.22–0.69 mg/L, which is lower compared with results of this study. However, if immobilization, which correlates closely with death, is considered as the endpoint, the values we identified are comparable with those observed by Blaise et al. [29]. The EC50 values for *T. fulvus* were included between 0.15 and 0.71 mg/L (Table 3), very similar to those reported by Blaise et al. [29]. Moreira et al. [50] reported the high sensitivity of oyster embryo-larval development, with EC50s of 0.15 mg/L after 24 h and 0.22 mg/L after 48 h exposure to Y.

Table 3. EC20 and EC50 (mg/L) with 95% C.I., of immobilization of *T. fulvus* exposed to REEs. Different letters within each column indicate significant differences ($p < 0.05$).

REEs	EC50	C.I. (95%)	EC20	C.I. (95%)
Scandium	0.57 cde	0.69–0.47	0.36 de	0.38–0.35
Yttrium	0.49 cd	0.67–0.36	0.20 b	0.23–0.18
Lanthanum	0.37 bc	0.51–0.27	0.17 b	0.19–0.16
Cerium	0.74 e	0.89–0.61	0.40 e	0.42–0.37
Neodynium	0.19 ab	0.28–0.12	0.06 a	0.06–0.05
Gadolinum	0.15 a	0.20–0.12	0.07 a	0.08–0.06
Dysprosium	0.60 de	0.75–0.48	0.28 c	0.31–0.26
Holmium	0.44 cd	0.58–0.34	0.19 b	0.21–0.17
Erbium	0.71 e	0.90–0.56	0.33 d	0.36–0.30

Not surprisingly, sensitivity to REEs differs greatly among organisms and among the endpoints used, making comparisons difficult. In a toxicity study of seven REEs (yttrium, lanthanum, cerium, neodymium, samarium, europium, and gadolinium) on three echinoderm species, Trifuoggi et al. [51] found that *Sphaerechinus granularis* had a significantly higher sensitivity to Y and La (EC50 of 0.042 mg/L and 0.008 mg/L, respectively) than *Arbacia lixula* and *Paracentrotus lividus*. Moreover, *P. lividus* embryos exposed to Gd showed higher sensitivity (EC50 = 0.031 mg/L) than *A. lixula* and *S. granularis*. Also, Martino et al. [52] observed different sensitivities to Gd in a study on the sea urchin embryonic development of the European (*Paracentrotus lividus* and *Arbacia lixula*) and Australian species (*Heliocidaris tuberculate* and *Centrostephanus rodgersii*), suggesting that it may be toxic to marine organisms even within the same taxonomic group.

Exposure to REEs resulted in concentration-dependent increases in mortality and immobilization for all REEs tested (Figures 1 and 2). Most of the REEs showed a significant acute toxic effect compared to the control, even at the lowest concentrations, confirming a toxicity that should not be underestimated. Concerning *T. fulvus* exposure to La and Gd, the effects (mortality) ranged between 20% (0.2 mg/L) and 93% (3.2 mg/L) for La, and between 13% (0.2 mg/L) and 97% (6.4 mg/L) for Gd. For Er, the mortality varied between 3% (0.2 mg/L) and 90% (6.4 mg/L). As regards immobilization, exposure to Gd showed the effect varying between 20% (0.1 mg/L) and 100% (1.6 mg/L) (Figures 1 and 2).

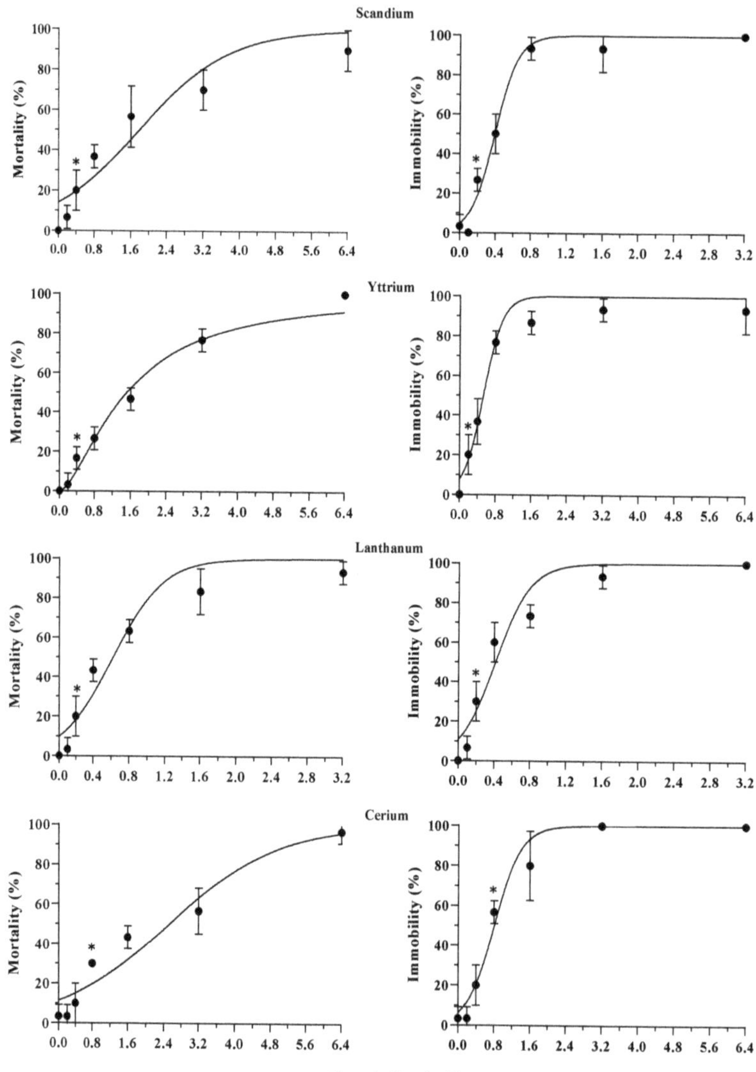

Figure 1. Concentration-response curves based on the mean (±s.d.; $n = 30$) mortality and immobilization rates of *T. fulvus* exposed to scandium, yttrium, lanthanum, and cerium. Asterisks indicate the first concentration causing significant differences to the control (Tukey's test, $p < 0.05$).

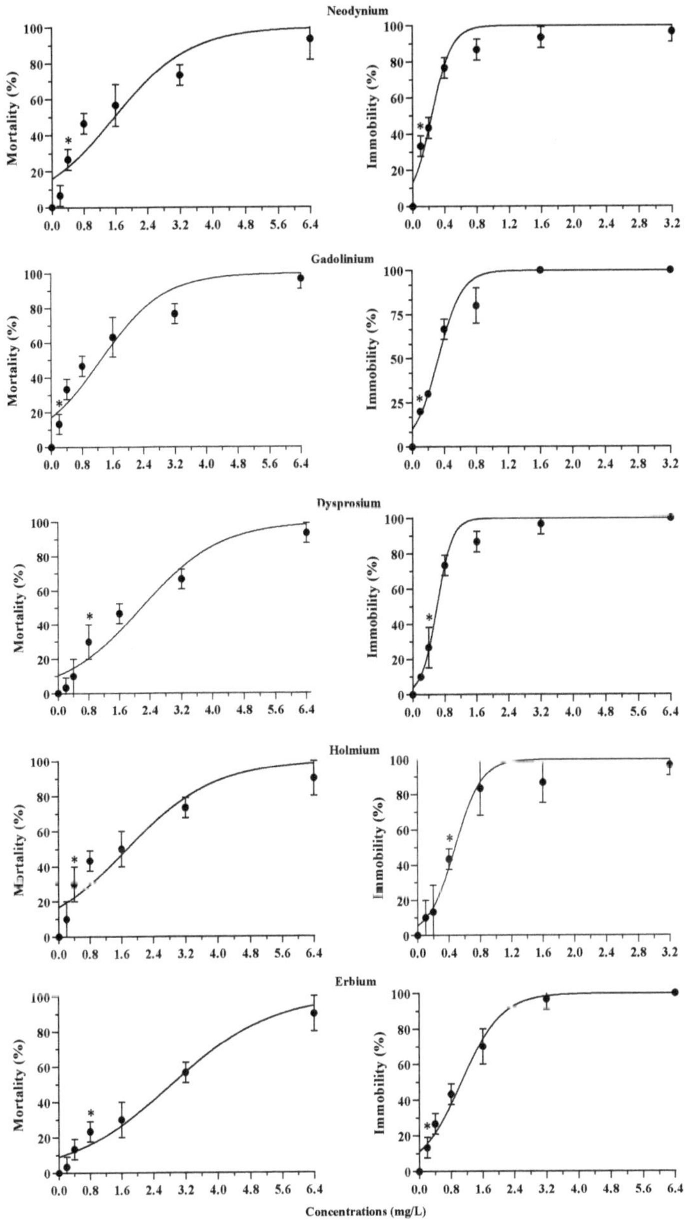

Figure 2. Concentration-response curves based on the mean (±s.d.; $n = 30$) mortality and immobilization rates of *T. fulvus* exposed to neodymium, gadolinium, dysprosium, holmium, and erbium. Asterisks indicate the lowest concentrations which show significant differences with the controls ($p < 0.05$).

Furthermore, all REEs negatively affected the larval development of *T. fulvus*. Exposure to all REEs resulted in a concentration-dependent decrease in the number of naupliar moults, which was significant compared to the control even at the lowest concentration tested (0.2 mg/L). (Figure 3, $p < 0.05$). At the end of the test, a total of 20–25 naupliar moults were found in the negative control. After exposure of nauplii to REEs, a significant slowdown of naupliar development was exhibited by all tested REEs. However, at the highest concentration of 6.4 mg/L, a 100% reduction in naupliar moults was observed only in copepods exposed to Y, displaying the most severe sublethal effect compared to the other REEs. This sub-lethal effect may be a potential harmful signal, affecting individual fitness and overall reproductive output, offering insight into the future health of a population.

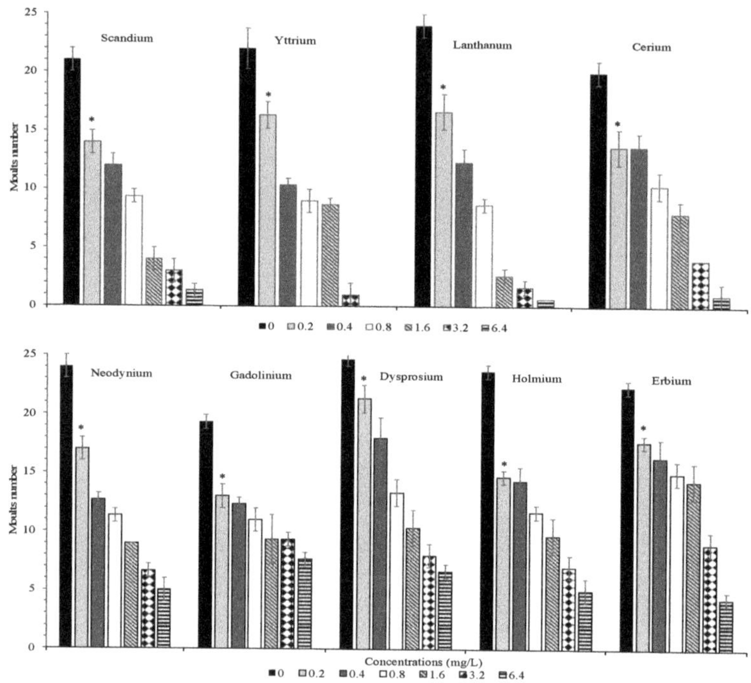

Figure 3. Moult numbers (mean ± s.d.) of *T. fulvus* exposed to different concentrations of REEs. Asterisks indicate the lowest concentrations which show significant differences with the controls ($p < 0.05$).

4. Conclusions

The continuous release of REEs into the marine environment poses a threat to marine species.

This study analyzed the acute and sub-lethal effects of nine REEs on the copepod *T. fulvus*. It is necessary to emphasize the importance of sub-lethal effects in ecotoxicological risk assessment, as reduced embryonic development and survival may pose a significant threat to adult recruitment and population health.

The results show more severe effects for La, Gd, and Nd on the copepod for all endpoints analyzed, while Er was the least toxic. Overall, a significant difference emerged between control copepods and those exposed to REEs even at the lowest concentrations. Although for some REEs, the LC50 and EC50 values obtained are outside the range of concentrations found in the natural environment, it must be considered that, given the role of REEs in our daily lives, their increasing demand, and consequent release into the environment, the results of this study can help to improve our understanding of the environmental impact of REEs and contribute to the definition of future guideline values and ecological risk calculations. In addition, further studies on chronic exposure to REE are needed to assess their effects on all life stages, as this could lead to significant changes.

Author Contributions: Conceptualization, E.P. and G.L.; methodology, F.B., A.G., M.T. and I.P.; software, F.B., A.G., M.T. and I.P.; validation, E.P., G.L., M.T. and I.P.; formal analysis, F.B., A.G., M.T. and I.P.; investigation, F.B., A.G. and I.P.; resources, G.L.; data curation, F.B., G.L. and M.T.; writing—original draft, F.B., E.P. and G.L.; writing—review and editing, E.P.; funding acquisition, E.P. All authors have read and agreed to the published version of the manuscript.

Funding: This research received no external funding.

Institutional Review Board Statement: Not applicable.

Informed Consent Statement: Not applicable.

Data Availability Statement: The original contributions presented in the study are included in the article, further inquiries can be directed to the corresponding authors.

Conflicts of Interest: The authors declare no conflicts of interest.

References

1. Xie, F.; Zhang, T.A.; Dreisinger, D.; Doyle, F. A critical review on solvent extraction of rare earths from aqueous solutions. *Miner. Eng.* **2014**, *56*, 10–28. [CrossRef]
2. Charalampides, G.; Vatalis, K.I.; Apostoplos, B.; Ploutarch-Nikolas, B. Rare earth elements: Industrial applications and economic dependency of Europe. *Procedia Econ. Financ.* **2015**, *24*, 126–135. [CrossRef]
3. Balaram, V. Rare earth elements: A review of applications, occurrence, exploration, analysis, recycling, and environmental impact. *Geosci. Front.* **2019**, *10*, 1285–1303. [CrossRef]
4. U.S. Geological Survey. Rare Earth Elements-Critical Resources for High Technology. Available online: http://geopubs.wr.usgs.gov/fact-sheet/fs087-02/ (accessed on 10 July 2024).
5. Massari, S.; Ruberti, M. Rare earth elements as critical raw materials: Focus on international markets and future strategies. *Resour. Policy* **2013**, *38*, 36–43. [CrossRef]
6. U.S. Geological Survey. *Mineral Commodity Summaries 2023*; U.S. Geological Survey: Reston, VA, USA, 2023; 201p, ISSN 0076-8952. [CrossRef]
7. Liu, W.S.; Guo, M.N.; Liu, C.; Yuan, M.; Chen, X.T.; Huot, H.; Zhao, C.-M.; Tang, Y.T.; Morel, J.L.; Qiu, R.L. Water, sediment and agricultural soil contamination from an ion-adsorption rare earth mining area. *Chemosphere* **2019**, *216*, 75–83. [CrossRef]
8. Li, X.; Chen, Z.; Chen, Z.; Zhang, Y. A human health risk assessment of rare earth elements in soil and vegetables from a mining area in Fujian Province, Southeast China. *Chemosphere* **2013**, *93*, 1240–1246. [CrossRef]
9. Zhuang, M.; Wang, L.; Wu, G.; Wang, K.; Jiang, X.; Liu, T.; Xiao, P.; Yu, L.; Jiang, Y.; Song, J.; et al. Health risk assessment of rare earth elements in cereals from mining area in Shandong, China. *Sci. Rep.* **2017**, *7*, 9772. [CrossRef]
10. Zhuang, M.; Zhao, J.; Li, S.; Liu, D.; Wang, K.; Xiao, P.; Yu, L.; Jiang, Y.; Song, J.; Zhou, J.; et al. Concentrations and health risk assessment of rare earth elements in vegetables from mining area in Shandong, China. *Chemosphere* **2017**, *168*, 578–582. [CrossRef]
11. Li, X.; Chen, Z.; Chen, Z. Distribution and fractionation of rare earth elements in soil–water system and human blood and hair from a mining area in southwest Fujian Province, China. *Environ. Earth Sci.* **2014**, *72*, 3599–3608. [CrossRef]
12. Khan, A.M.; Behkami, S.; Yusoff, I.; Md Zain, S.B.; Bakar, N.K.A.; Bakar, A.F.A.; Alias, Y. Geochemical characteristics of rare earth elements in different types of soil: A chemometric approach. *Chemosphere* **2017**, *184*, 673–678. [CrossRef]
13. Rogowska, J.; Olkowska, E.; Ratajczyk, W.; Wolsk, L. Gadolinium as a new emerging contaminant of aquatic environments. *Environ. Toxicol. Chem.* **2018**, *37*, 1523–1534. [CrossRef] [PubMed]
14. MacMillan, G.A.; Chételat, J.; Heath, J.P.; Mickpegak, R.; Amyot, M. Rare earth elements in freshwater, marine, and terrestrial ecosystems in the eastern. *Environ. Sci. Process. Impacts* **2017**, *19*, 1336–1345. [CrossRef]
15. Romero-Freire, A.; Minguez, L.; Pelletier, M.; Cayer, A.; Caillet, C.; Devin, S.; Gross, E.M.; Guérold, F.; Pain-Devin, S.; Vignati, D.A.L.; et al. Assessment of baseline ecotoxicity of sediments from a prospective mining area enriched in light rare earth elements. *Sci. Total Environ.* **2018**, *612*, 831–839. [CrossRef]
16. Bau, M.; Dulski, P. Anthropogenic origin of positive gadolinium anomalies in river waters. *Earth Planet. Sci. Lett.* **1996**, *143*, 245–255. [CrossRef]
17. Birka, M.; Wehe, C.A.; Hachmoller, O.; Sperling, M. Tracing gadolinium-based contrast agents from surface water drinking water by means of speciation analysis. *J. Chromatogr.* **2016**, *1440*, 105–111. [CrossRef]
18. Raju, C.S.K.; Cossmer, A.; Scharf, H.; Panne, U.; Lück, D. Speciation of gadolinium based MRI contrast agents in environmental water samples using hydrophilic interaction chromatography hyphenated with inductively coupled plasma mass spectrometry. *J. Anal. At. Spectrom.* **2010**, *25*, 55–61. [CrossRef]
19. Merschel, G.; Bau, M. Rare earth elements in the aragonitic shell of freshwater mussel *Corbicula fluminea* and the bioavailability of anthropogenic lanthanum, samarium and gadolinium in river water. *Sci. Total Environ.* **2015**, *533*, 91–101. [CrossRef] [PubMed]
20. Olías, M.; Cerón, J.C.; Fernández, I.; De La Rosa, J. Distribution of rare earth elements in an alluvial aquifer affected by acid mine drainage: The Guadiamar aquifer (SW Spain). *Environ. Pollut.* **2005**, *135*, 53–64. [CrossRef]

21. Sultan, K.; Shazili, N.A. Rare earth elements in tropical surface water, soil and sediments of the Terengganu River Basin, Malaysia. *J. Rare Earths* **2009**, *27*, 1072–1078. [CrossRef]
22. Migaszewski, Z.M.; Gałuszka, A. The characteristics, occurrence, and geochemical behavior of rare earth elements in the environment: A review. *Crit. Rev. Environ. Sci. Technol.* **2015**, *45*, 429–471. [CrossRef]
23. Uchida, S.; Tagami, K.; Tabei, K.; Hirai, I. Concentrations of REEs, Th and U in river waters collected in Japan. *J. Alloys Compd.* **2006**, *408*, 525–528. [CrossRef]
24. González, V.; Vignati, D.A.; Pons, M.N.; Montarges-Pelletier, E.; Bojic, C.; Giamberini, L. Lanthanide ecotoxicity: First attempt to measure environmental risk for aquatic organisms. *Environ. Pollut.* **2015**, *199*, 139–147. [CrossRef]
25. Lachaux, N.; Otero-Fariña, A.; Minguez, L.; Sohm, B.; Rétif, J.; Châtel, A.; Poirier, L.; Devin, S.; Pain-Devin, S.; Gross, E.M.; et al. Fate, subcellular distribution and biological effects of rare earth elements in a freshwater bivalve under complex exposure. *Sci. Total Environ.* **2023**, *905*, 167302. [CrossRef]
26. Mayfield, D.; Fairbrother, A. Examination of rare earth element concentration patterns in freshwater fish tissues. *Chemosphere* **2014**, *120*, 68–74. [CrossRef] [PubMed]
27. Blinova, I.; Muna, M.; Heinlaan, M.; Lukjanova, A.; Kahru, A. Potential Hazard of Lanthanides and Lanthanide-Based Nanoparticles to Aquatic Ecosystems: Data Gaps, Challenges and Future Research Needs Derived from Bibliometric Analysis. *Nanomaterials* **2020**, *10*, 328. [CrossRef] [PubMed]
28. Malhotra, N.; Hsu, H.-S.; Liang, S.-T.; Roldan, M.J.M.; Lee, J.-S.; Ger, T.-R.; Hsiao, C.-D. An Updated Review of Toxicity Effect of the Rare Earth Elements (REEs) on Aquatic Organisms. *Animals* **2020**, *10*, 1663. [CrossRef]
29. Blaise, C.; Gagné, F.; Harwood, M.; Quinn, B.; Hanana, H. Ecotoxicity responses of the freshwater cnidarian *Hydra attenuata* to 11 rare earth elements. *Ecotoxicol. Environ. Saf.* **2018**, *163*, 486–491. [CrossRef]
30. Blinova, I.; Lukjanova, A.; Muna, M.; Vija, H.; Kahru, A. Evaluation of the potential hazard of lanthanides to freshwater microcrustaceans. *Sci. Total Environ.* **2018**, *642*, 1100–1107. [CrossRef]
31. Cui, J.; Zhang, Z.; Bai, W.; Zhang, L.; He, X.; Ma, Y.; Liu, Y.; Chai, Z. Effects of rare earth elements la and yb on the morphological and functional development of zebrafish embryos. *J. Environ. Sci.* **2012**, *24*, 209–213. [CrossRef]
32. Mestre, N.C.; Sousa, V.S.; Rocha, T.L. Ecotoxicity of rare earths in the marine mussel *Mytilus galloprovincialis* and a preliminary approach to assess environmental risk. *Ecotoxicology* **2019**, *28*, 294–301. [CrossRef]
33. Freitas, R.; Costa, S.; Cardoso, C.E.; Morais, T.; Moleiro, P.; Matias, A.C.; Pereira, A.F.; Machado, J.; Correia, B.; Pinheiro, D.; et al. Toxicological effects of the rare earth element neodymium in *Mytilus galloprovincialis*. *Chemosphere* **2020**, *244*, 125457. [CrossRef] [PubMed]
34. Pagano, G.; Guida, M.; Siciliano, A.; Oral, R.; Koçbas, F.; Palumbo, A.; Castellano, I.; Migliaccio, O.; Thomas, P.J.; Trifuoggi, M. Comparative toxicity of selected rare earth elements: Sea urchin embryogenesis and fertilization damage with redo and cytogenetic effect. *Environ. Res.* **2016**, *147*, 453–460. [CrossRef] [PubMed]
35. Siciliano, A.; Guida, M.; Serafini, S.; Micillo, M.; Galdiero, E.; Carfagna, S.; Salbitani, G.; Tommasi, F.; Lofrano, G.; Suarez, E.G.P.; et al. Long-term multi-endpoint exposure of the microalga *Raphidocelis subcapitata* to lanthanum and cerium. *Sci. Total Environ.* **2021**, *790*, 148229. [CrossRef] [PubMed]
36. Sun, D.; He, N.; Chen, Q.; Duan, S. Effects of Lanthanum on the photosystem II energy fluxes and antioxidant system of *Chlorella vulgaris* and *Phaeodactylum tricornutum*. *Int. J. Environ. Res. Public Health* **2019**, *16*, 2242. [CrossRef]
37. Tai, P.; Zhao, Q.; Su, D.; Li, P.; Stagnitti, F. Biological toxicity of lanthanide elements on algae. *Chemosphere* **2010**, *80*, 1031–1035. [CrossRef]
38. Faraponova, O.; Giacco, E.; Biandolino, F.; Prato, E.; Del Prete, F.; Valenti, A.; Sarcina, S.; Pasteris, A.; Montecavalli, A.; Comin, S.; et al. *Tigriopus fulvus:* The interlaboratory comparison of the acute toxicity test. *Ecotoxicol. Environ. Saf.* **2016**, *124*, 309–314. [CrossRef]
39. UNICHIM 2396: 2014. Qualità Dell'acqua—Determinazione Della Tossicità Letale a 24 h, 48 h e 96 h di Esposizione con Naupli di Tigriopus Fulvus (Fischer, 1860). (Crustacea:Copepoda). 2014. Available online: https://pubblicazioni.unichim.it/metodi (accessed on 11 July 2024).
40. Prato, E.; Biandolino, F.; Parlapiano, I.; Grattagliano, A.; Rotolo, F.; Buttino, I. Historical control data of ecotoxicological test with the copepod *Tigriopus fulvus*. *Chem. Ecol.* **2023**, *39*, 881–893. [CrossRef]
41. Morel, E.; Cui, L.; Zerges, W.; Wilkinson, K.J. Mixtures of rare earth elements show antagonistic interactions in *Chlamydomonas reinhardtii*. *Environ. Pollut.* **2021**, *287*, 117594. [CrossRef]
42. Huang, Z.; Gao, N.; Zhang, S.; Xing, J.; Hou, J. Investigating the toxically homogenous effects of three lanthanides on zebrafish. *Comp. Biochem. Physiol. Part C Toxicol. Pharmacol.* **2022**, *253*, 109251. [CrossRef]
43. Lachaux, N.; Cossu-Leguille, C.; Poirier, L.; Gross, E.M.; Giamberini, L. Integrated environmental risk assessment of rare earth elements mixture on aquatic ecosystems. *Front. Environ. Sci.* **2022**, *10*, 974191. [CrossRef]
44. Lin, Y.T.; Liu, R.X.; Audira, G.; Suryanto, M.E.; Roldan, M.J.M.; Lee, J.S.; Ger, T.R.; Hsiao, C.D. Lanthanides toxicity in zebrafish embryos are correlated to their atomic number. *Toxics* **2022**, *10*, 336. [CrossRef] [PubMed]
45. Gu, Y.G.; Wang, Y.S.; Jordan, R.W.; Su, H.; Jiang, S.J. Probabilistic ecotoxicological risk assessment of heavy metal and rare earth element mixtures in aquatic biota using the DGT technique in coastal sediments. *Chemosphere* **2023**, *329*, 138592. [CrossRef] [PubMed]

46. Siciliano, A.; Sabatino, M.; Paone, A.; Padilla Suarez, E.G.; Toscanesi, M.; Brouziotis, A.A.; Gambino, E.; Saviano, L.; Trifuoggi, M.; Guida, M.; et al. A first attempt to evaluate the toxicity to *Phaeodactylum tricornutum* Bohlin exposed to rare earth elements. *Front. Environ. Sci.* **2022**, *10*, 957943. [CrossRef]
47. Markich, S.J.; Hall, J.P.; Dorsman, J.M.; Brown, P.L. Toxicity of rare earth elements (REEs) to marine organisms: Using species sensitivity distributions to establish water quality guidelines for protecting marine life. *Environ. Res.* **2024**, *261*, 119708. [CrossRef]
48. Borgmann, U.; Couillard, Y.; Doyle, P.; Dixon, D.G. Toxicity of sixty-three metals and metalloids to *Hyalella azteca* at two levels of water hardness. *Environ. Toxicol. Chem.* **2005**, *24*, 641–652. [CrossRef]
49. Manusadžianas, L.; Vitkus, R.; Gylytė, B.; Cimmperman, R.; Džiugelis, M.; Karitonas, R.; Sadauskas, K. Ecotoxicity Responses of the Macrophyte Algae *Nitellopsis obtusa* and freshwater Crustacean *Thamnocephalus platyurus* to 12 Rare Earth Elements. *Sustainability* **2020**, *12*, 7130. [CrossRef]
50. Moreira, A.; Bruno, H.; Leite, C.; Libralato, G.; Pereira, E.; Freitas, R. Potential impacts of lanthanum and yttrium through embryotoxicity assays with *Crassostrea gigas*. *Ecol. Indic.* **2020**, *108*, 105687. [CrossRef]
51. Trifuoggi, M.; Pagano, G.; Guida, M.; Palumbo, A.; Siciliano, A.; Gravina, M.; Lyons, D.M.; Burić, P.; Levak, M.; Philippe, J.; et al. Comparative toxicity of seven rare earth elements in sea urchin early life stages. *Environ. Sci. Pollut. Res.* **2017**, *24*, 20803–20810. [CrossRef]
52. Martino, C.; Bonaventura, R.; Byrne, M.; Roccheri, M.; Matranga, V. Effects of exposure to gadolinium on the development of geographically and phylogenetically distant sea urchin species. *Mar. Environ. Res.* **2017**, *128*, 98–106. [CrossRef]

Journal of *Xenobiotics*

MDPI

Article

Assessment of Chicken Fecal Contamination Using Microbial Source Tracking (MST) and Environmental DNA (eDNA) Profiling in Silway River, Philippines

Lonny Mar Opog [1], Joan Cecilia Casila [2,*], Rubenito Lampayan [2], Marisa Sobremisana [3], Abriel Bulasag [4], Katsuhide Yokoyama [5] and Soufiane Haddout [6]

1 Agricultural and Biosystems Engineering Department, College of Agriculture, Mindanao State University, General Santos City 9500, Philippines; lonnymar.opog@msugensan.edu.ph

2 Land and Water Resources Engineering Division, Institute of Agricultural and Biosystems Engineering, College of Engineering and Agro-Industrial Technology, University of the Philippines Los Baños, Laguna 4031, Philippines; rmlampayan@up.edu.ph

3 School of Environmental Science and Management, University of the Philippines Los Baños, Laguna 4031, Philippines

4 University of the Philippines Rural High School, College of Arts and Sciences, University of the Philippines Los Baños, Laguna 4033, Philippines; asbulasag@up.edu.ph

5 Department of Civil and Environmental Engineering, Tokyo Metropolitan University, Hachioji 192-0397, Tokyo, Japan; k-yoko@tmu.ac.jp

6 Department of Physics, Faculty of Science, Ibn Tofail University, Kenitra 14000, Morocco; soufian.haddout@gmail.com

* Correspondence: jmcatubigcasila@up.edu.ph

Abstract: The Silway River has historically failed to meet safe fecal coliform levels due to improper waste disposal. The river mouth is located in General Santos City, the tuna capital of the Philippines and a leading producer of hogs, cattle, and poultry. The buildup of contaminants due to direct discharge of waste from chicken farms and existing water quality conditions has led to higher fecal matter in the Silway River. While there were technical reports in the early 2000s about poultry farming, this is the first study where fecal coliform from poultry farming was detected in the Silway River using highly sensitive protocols like qPCR. This study characterized the effect of flow velocity and physicochemical water quality parameters on chicken fecal contamination. Gene markers such as Ckmito and ND5-CD were used to detect and quantify poultry manure contamination through microbial source tracking (MST) and environmental DNA (eDNA) profiling. The results of this study showed the presence of chicken fecal bacteria in all stations along the Silway River. The results revealed that normal levels of water quality parameters such as temperature, pH, and high TSS concentrations create favorable conditions for chicken fecal coliforms to thrive. Multiple regression analysis showed that flow velocity and DO significantly affect chicken fecal contamination. A lower cycle threshold (Ct) value indicated higher concentration of the marker ND5-CD, which means higher fecal contamination. It was found that there was an inverse relationship between the Ct value and both velocity ($R^2 = 0.55$, $p = 0.01$) and DO ($R^2 = 0.98$, $p = 0.2$), suggesting that low flow velocity and low DO can lead to higher fecal contamination. Findings of fecal contamination could negatively impact water resources, the health of nearby residents, and surrounding farms and industries, as well as the health and growth of fish.

Keywords: chicken fecal contamination; microbial source tracking; eDNA profiling; fecal contamination; flow velocity; water quality; Silway River

1. Introduction

Domestic chickens are common around the world, especially in the Philippines. They are one of the most essential sub-sectors of the Philippines' agricultural industry. Chickens are typically the primary source of income in rural nations [1]. Consumers have a much greater preference for chicken meat, which has resulted in its being one of the most consumed meats due to its affordability. Among all the animals raised primarily for human consumption, one of the most consumed were chickens, which significantly contribute to the country's livestock production volume.

According to the Philippine Statistics Authority's situation report for the period October to December 2022, total chicken production was estimated at 494.12 thousand metric tons (live weight). Concurrently, the national chicken inventory was reported to be approximately 188.48 million fowls. Among the 18 regions in the Philippines, Region 12, also known as SOCCSKSARGEN, is derived from the provinces and cities that make up the region—South Cotabato, Cotabato, Sultan Kudarat, Sarangani, and General Santos City (Figure 1). The region ranks seventh among the major chicken-producing areas in the Philippines, contributing approximately 22,249 metric tons (live weight) to total chicken production. Additionally, it accounts for an estimated 10.37 million fowls in the national chicken inventory [2]. According to the Philippine Bureau of Animal Industry report, as of December 2022, there were a total of one hundred and eighty (180) registered poultry farms in South Cotabato.

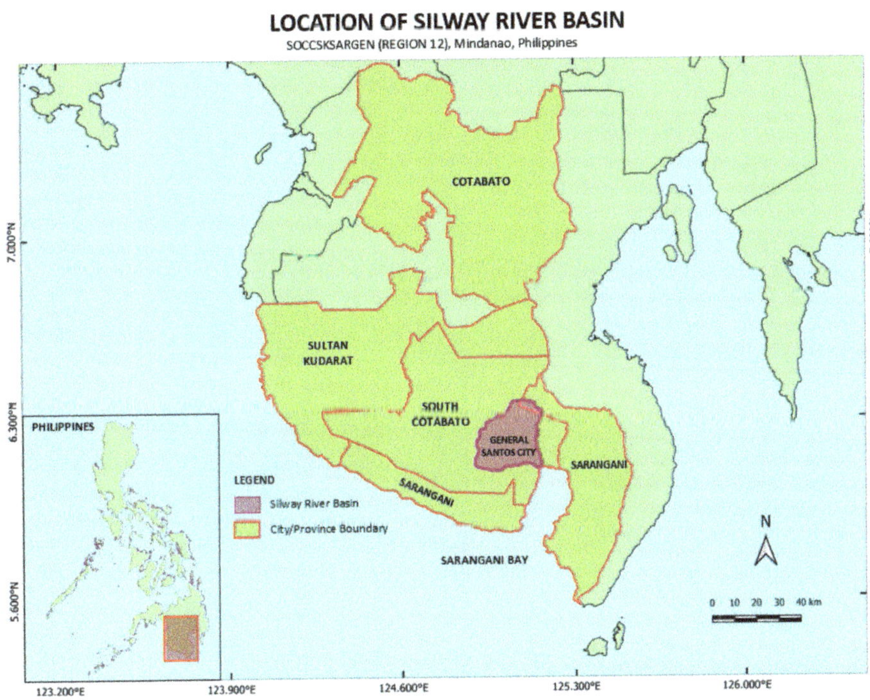

Figure 1. Location of Silway River basin in Southern Mindanao, Philippines.

The increasing demand for poultry products has raised waste management as a prominent issue within environmental considerations. One of the major sources of waste that contribute to environmental pollution is the chicken industry [3,4]. According to the local government, one of the issues experienced in this industry is the unregulated waste disposal methods and practices of poultry owners, which has resulted in various issues, particularly for the environment [3]. Chicken feces are released into the environment and receiving bodies of water as a result of commercial and backyard poultry raising, particularly during storms and floods [5].

Poultry fecal contamination poses sanitary risks by introducing pathogens like Escherichia coli, Salmonella spp., Campylobacter spp., and avian influenza viruses into water systems. Coliform bacteria, while useful indicators, do not directly reflect pathogen presence [6]. These microorganisms can cause gastrointestinal illnesses, systemic infections, and zoonotic outbreaks. Nutrient enrichment from poultry waste also promotes eutrophication, prolonging pathogen survival and fostering harmful algal blooms [7]. Addressing these risks is vital to safeguard water quality and public health.

According to research by the Department of Environment and Natural Resources, there are 180 out of 421 rivers in the country that are dirty and polluted and may soon be declared "biologically dead" [8], while 50 rivers are already considered "biologically dead". Due to this situation, the local government agency identified several rivers in the country as Water Quality Management Areas in accordance with the Philippine Clean Water Act of 2004. This study aligns with the United Nations Sustainable Development Goal (SDG) 6, "Clean Water and sanitation", which aims to ensure the availability and sustainable management of water and sanitation for all. With respect to SDG 14, "Life Below Water", this work provides a novel tool for detecting and managing water pollution, leading to improved water quality especially for Sarangani Bay. Sarangani Bay is a significant source of income for the Sarangani province, and General Santos City, the tuna capital of the Philippines, drains to this water body. By employing an innovative approach to environmental monitoring and management, this work enhances water quality by reducing pollution, minimizing fecal contamination, and significantly increasing the availability of safe water reuse on a global scale.

Traditionally, fecal indicator bacteria (FIB) have been used to identify and count the presence of fecal water pollution, including fecal coliforms and enterococci [5], but due to a lack of host specificity related to genetic markers shared universally across a diverse range of hosts, FIB analysis is unable to provide information on possible sources of contamination [9]. In this study, a microbial source tracking (MST) method was used by focusing on host-specific mitochondrial DNA (mtDNA) of chicken to determine the fecal contamination in the Silway River. MST refers to a set of methodologies used to identify and, in some cases, quantify the primary sources of fecal contamination in various water bodies, including drinking water, groundwater, and recreational waters. These methods are classified into library-dependent and library-independent approaches, each designed to trace contamination to its origin [10].

Environmental DNA (eDNA) was used in the Philippines for other applications like metabarcoding and fish visual census [11]. In the Laguna Lake, Philippines, MST using rep-PCR markers was used for fecal contamination studies [12,13]. In other countries, microbial and chemical source tracking markers were used to monitor fecal contamination at drinking water intakes [14]. In China [5] and Virginia, USA [15], eDNA and MST were used for quantitative detection of domestic poultry feces and human and poultry fecal contamination, respectively.

In 2010, the Sustainable Sanitation for East Asia (SuSEA) Philippines component developed the Water Pollution Prevention and Control Program (WPPCP) with the intended result of lowering the biochemical oxygen demand (BOD) loading of the Silway River in South Cotabato by 10% at the end of the year [16]. The program was developed with the desire to support the designation of the Silway River's WQMA. The fact that the Silway River was declared one of the Water Quality Management Areas (WQMAs) implies that the river's current state demands extensive protection and management by the local government and stakeholders. On 28 November 2022, the Green City Action Plan (GCAP) for General Santos City was published. It established a new target of 30% reduction on BOD load and fecal coliform in Sarangani Bay by 2040 (from the 2018 level). This 20% increase in the reduction goal indicates that the water quality deteriorated over the past years [17].

The novelty of this study lies in the fact that it is the first local study in the Philippines about the application of host-specific detection and quantification of chicken fecal contamination in a river system. This study could also be used in determining chicken fecal contamination in soil samples. This study would open new research, water quality monitoring technique, and potential local government projects to address chicken fecal contamination not only in rivers but also in various aquatic and terrestrial environments, especially in regions with high total chicken meat production in the country. Lastly, the results of this study could help in the assessment of the water quality of Silway River stations that are not covered by the Silway River Water Quality Management Area.

This study aims to examine the level of chicken fecal contamination in the Silway River, Philippines, on a spatial scale. Several critical water quality parameters that are crucial to monitoring and environmental management were also assessed. Specifically, this study characterized flow velocity and physicochemical water quality parameters; utilized two gene markers, Ckmito and ND5-CD, to detect and quantify poultry manure contamination; evaluated chicken fecal contamination using MST and eDNA profiling; and assessed the relationship between chicken fecal contamination and physicochemical water quality parameters. The methodology employed in this study can be utilized for rivers and water bodies worldwide, providing a valuable tool for environmental monitoring and management across various contexts and regions.

2. Materials and Methods

2.1. Site Description

One of the rivers under protection in the Philippines is the Silway River in Mindanao. The Silway River estuary flows through a densely populated residential area with commercial establishments [18]. The Silway River is located in the southern part of South Cotabato province and the western part of Sarangani province (Figure 1). The Silway River WQMA comprises the municipalities of Polomolok, Tupi, and T'boli and General Santos City. Subsequently, these areas' wastewater discharges drain into the Silway River.

The river basin falls within fifty-nine barangay jurisdictions, distributed in the four abovementioned cities or municipalities, and drains into Sarangani Bay [19]. The water quality of the Silway river directly impacts the bay [17], which is a protected seascape and a renowned spawning habitat for tuna fingerlings. General Santos City is a highly urbanized city in South Cotabato. The Silway River mouth is located in this city, which is known as the tuna capital of the Philippines with a GDP that is highly dependent on fish catch. Agriculture and fisheries are the main drivers of General Santos City's economy.

2.2. Water Sampling Materials and Data Collection Equipment

Different materials and equipment were used in this study. An electromagnetic current meter (INFINITY-EM, JFE Advantech, Nishinomiya, Japan) with a data logger and built-in compass was used to measure the flow velocity. A YSI ProDSS multiparameter was used to identify the actual coordinates of the sampling stations and measure DO, turbidity, and temperature. A leveling rod was used to determine the depth of the river. A digital pH meter was used to measure the pH.

2.3. Preliminary Fieldwork Preparation

A preliminary reconnaissance survey was conducted to identify the location of the sampling stations and nearby poultry farms and backyard poultries in Silway River. The initial coordinates of the stations were identified using Google Earth Pro. The selection of water sampling stations was determined based on the proximity to poultry farms and local communities engaged in backyard poultry raising and family poultry. Additionally, some stations were selected in alignment with the Silway River Water Quality Management Area (WQMA) stations. Figure 2 shows the map of the sources of chicken fecal contamination.

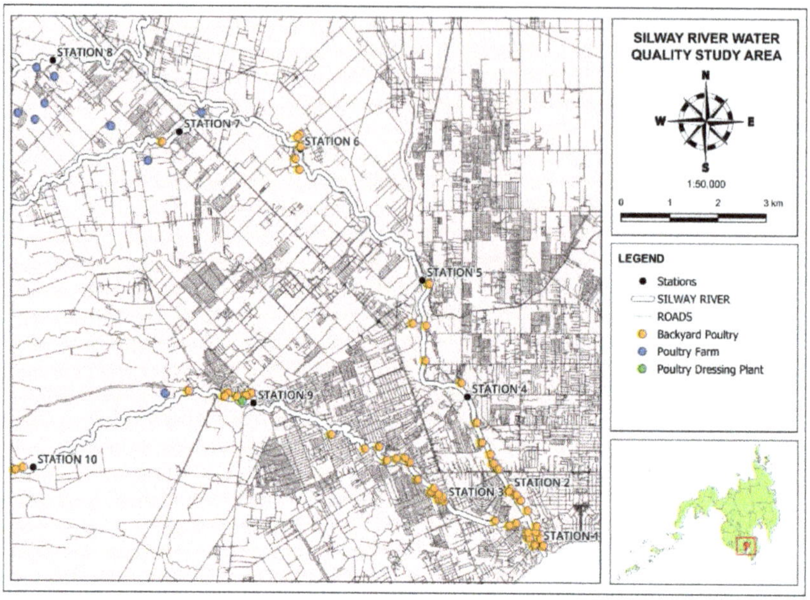

Figure 2. Sources of chicken fecal contamination and sampling locations along the Silway River tributaries in South Cotabato and General Santos City. The red square shows the location of Silway River basin.

2.4. Methods for Water Sampling, Data Gathering, and Laboratory and Data Analysis

Figure 3 shows a flow chart with a brief procedural illustration of the study. Flow velocity, actual stations coordinates, water quality parameters, and water samples for laboratory tests were gathered simultaneously.

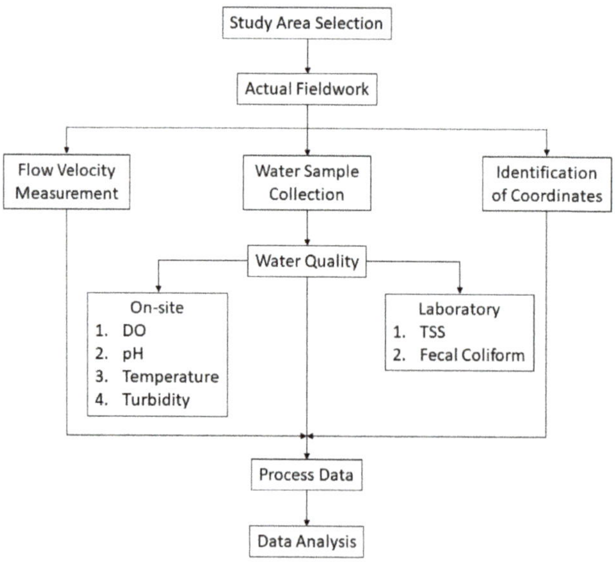

Figure 3. Procedural flow chart.

2.4.1. Study Area Selection

The study area is located in the province of South Cotabato, a region in Southern Mindanao. The location of each sampling station was determined using an electromagnetic current meter with a data logger and built-in compass. The catchment area of the Silway River has a total of 56,280 hectares composed of different municipalities including Polomolok, with most of the portions and parts of T'boli, Tupi, and General Santos City. It is drained by several streams and tributaries that make up the 28 km Silway River [16]. Figure 4 shows the different sampling stations of the Silway River.

Figure 4. Images of the sampling locations: (**A**) Station 1, (**B**) Station 2, (**C**) Station 3, (**D**) Station 4, (**E**) Station 5, (**F**) Station 6, (**G**) Station 7, (**H**) Station 8, (**I**) Station 9, (**J**) Station 10.

2.4.2. Water Sample Collection and Data Gathering

The study was conducted on 14 April 2023 from 8:00 a.m. to 4:00 p.m. when the weather was sunny and clear. A total of 10 water samples were collected for this study from the sampling stations of Silway River, province of South Cotabato, including the different barangays of General Santos City namely, Labangal, Sinawal, City Heights, and San Isidro, and barangays of Polomolok including Silway 7 and Silway 8. Table 1 shows a list of the sampling stations and their descriptions.

All equipment and materials were rinsed with distilled water. Each sampling bottle was labeled with station number from one (1) to ten (10). Flow velocity was measured using an electromagnetic current meter (INFINITY-EM, JFE Advantech), while the pH was measured using a digital pH meter immediately after the samples were collected in each station. The water quality parameters such as dissolved oxygen (DO), temperature, and turbidity were measured using a multiparameter water quality meter (YSI ProDSS, Xylem Analytics, OH, USA) with an optical luminescence sensor for dissolved oxygen, thermistor sensor for temperature, and optical 90-degree scatter sensor for turbidity. Water quality parameter measurements were performed at a depth of 0.1 m in each station. The width of the river varies in every station, so it was divided into three sections to measure the flow velocity and depth, and these sections were averaged later for the profile distribution of velocity and depth.

Table 1. Silway River sampling stations.

Station	Latitude	Longitude	Station Description
1	6.1058	125.1640	Approx. 300 m from the Mouth of Silway River GSC
2	6.1154	125.1585	Approx 435 m from Silway Bridge along Digos-Makar Rd. at Purok Palen Brgy. Labangal GSC
3	6.1136	125.1462	Approx. 340 m from Sinawal Bridge along Makar—Gensan Rd. at Purok San Vicente Brgy. Labangal GSC
4	6.1321	125.1505	Foot of bridge along Yumang St. Purok Sta. Teresita Rd. City Heights GSC
5	6.1531	125.1420	Foot of Upper Silway Bridge along GSC Circumferential Rd. at Brgy. San Isidro GSC
6	6.1765	125.1190	Foot of Silway 7 Bridge along Silway 7—Klinan 6 Rd. at Silway 7 Polomolok
7	6.1795	125.0966	Approx. 250 m from Matin-ao Bridge of Purok Riverside Matin-ao Silway 8 Polomolok
8	6.1923	125.0730	Downstream portion of Silway River at Purok Upper Matin-ao Silway 8 Polomolok
9	6.1308	125.1103	Foot of Upper Sinawal Bridge along GSC Circumferential Rd. at Purok Cabuay Brgy. Sinawal GSC
10	6.11913	125.06902	Upstream portion of Silway River at Purok Kintay Brgy. Sinawal GSC

2.4.3. Laboratory Analysis

Water samples for laboratory tests were collected from the middle section of the river. A total of 1500 mL of the water sample was collected in each station, 500 mL for total suspended solid (TSS) tests, and 1000 mL for PCR and qPCR tests for chicken fecal coliform. After collection, water samples were properly sealed to prevent contamination and stored immediately in the insulated ice container to keep the trace elements. The gathered water samples were transported to the laboratories within 24 to 48 h from collection. The water samples for TSS were sent to the Mindanao State University—GSC Water Laboratory—while the water samples for fecal coliform analysis were sent to the Philippine Genome Center Mindanao Omics Laboratory.

The TSS analysis followed the EPA (1983) Method 160.2 (Gravimetric, Dried at 103–105 °C). Total suspended solid measurement for each station was obtained by filtration and gravimetric methods using a standard glass fiber filter paper, oven dryer, analytical weighing scale, and vacuum filtration. Initially, the filter paper was weighed to identify the initial weight. Then, each water sample was poured into the vacuum filtration set up. After allowing the water to be drained, the residue left on the filter paper was oven dried at a constant temperature of 105 °C for 48 h. Lastly, the filter paper was weighed to obtain the final weight, and the value for total suspended solids was determined using the following formula:

$$TSS\left(\frac{mg}{L}\right) = \frac{W_f - W_i}{V} \times 1000000 \quad (1)$$

where W_i = initial weight (g), W_f = final weight (g), and V = volume (mL).

For fecal coliform analysis, 1000 mL of water samples were processed by the Philippine Genome Center Mindanao Omics Laboratory Services. For the MST method, Ckmito and ND5-CD primers and probes were used for the PCR and qPCR tests. The nucleotide sequences were retrieved from the study conducted by [5,20]. Minor groove binder (MGB) probe and primers were synthesized by Noveaulab. The oligonucleotide sequences used are shown in Table 2.

Table 2. Oligonucleotide primers and probes used for validation of MST qPCR and PCR assays.

Assay	Primer or Probe	Concentration in Final Reaction Mix	Oligonucleotide Sequence (5′-3′)	Product Size (bp)	Annealing T (°C)	Target Gene	Target Host
ND5-CD (TaqMan)	ND5-F ND5-R ND5-P	900 nM 900 nM 250 nM	ACCTCCCCCAACTAGC TTGCCAATGGTTAGGCAGGAG (6-FAM [1]) TCAACCCATGCCTTCTT (NFQ-MGB [2])	172	60	Mitochondrial genes: NADH dehydrogenase subunit 5 (ND5)	Poultry (Chicken and duck)
Ckmito1 (single PCR)	Ckmito1-G Ckmito1-D	200 nM 200 nM	ACCCTATTTGACTCCCTCAA ATGTCGACCAGGGGTTTATG	565	55	Mitochondrial 16S rRNA gene	Poultry (Chicken)
CkmitoN1 (Nested PCR)	CkmitoN1-G CkmitoN1-D	200 nM 200 nM	CCCCCACACTAACAAGCAAT GGTTGTAAGGTGGTCGTGAT	381	55	Mitochondrial 16S rRNA gene	Poultry (Chicken)

[1] 6-FAM: 6-Carboxy-Fluorescein. [2] NFQ-MGB: Non-Fluorescent Quencher-Minor Groove Binder.

First, environmental DNA (eDNA) was extracted from the water samples using the Qiagen PowerSoil Pro Kit (Qiagen, Hilden, Germany), adhering to the manufacturer's instructions. Then, the concentration of the extracted eDNA in each station was measured using the spectrophotometer. After determining the concentration of the eDNA in every station, Agarose Gel Electrophoresis was performed with 1% Agarose Gel concentration and electrophoresed for 60 min at 100 V. For the detection of chicken fecal contamination, the samples were amplified using the Ckmito1 forward and reverse primer for single PCR with 20 μL of reaction volume and annealing temperature of 55 °C and were run in a conventional PCR machine. Gel concentration was 2% and electrophoresed for 90 min at 100 V. The same process was performed for the nested PCR test using CkmitoN1 forward and reverse primer. The amplified products were obtained using the primers Ckmito 1-G/D, and then the positive products were used as DNA templates when the primers CkmitoN1-G/D were applied. The markers Ckmito1 and CkmitoN1 were combined and collectively referred to as the Ckmito marker. To quantify chicken fecal coliforms, a TaqMan assay targeting the ND5 gene was conducted using a real-time qPCR system (Applied Biosystems, Foster City, CA, USA). Positive products from the nested PCR amplification were used as the DNA template to measure the cycle threshold (Ct) value.

There were missing values for dissolved oxygen and turbidity that were observed from the report of the results and were later revealed as a result of instrumental error during the collection of samples in station 4. Although there were missing values, the researcher was able to estimate the turbidity measurement for station 4 using the conversion Equation (2) generated by performing linear regression using the linear relationship between analyzed suspended solids and turbidity from nine stations. Figure 5 shows the analysis for linear regression between turbidity (FNU) and analyzed TSS (mg/L).

$$Turbidity = 2.3061x \quad (2)$$

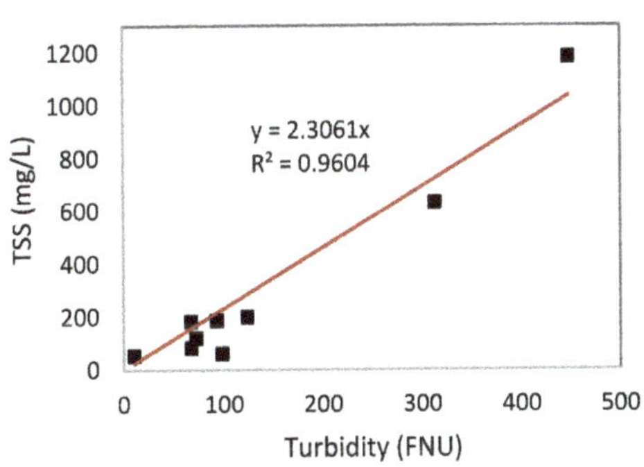

Figure 5. Regression analysis between turbidity (FNU) and analyzed total suspended solids (mg/L).

2.5. Data Mapping of Chicken Fecal Contamination and Physicochemical Water Quality Parameters Results

The coordinates of each sampling station were identified using the built-in compass in the electromagnetic current meter (INFINITY-EM, JFE Advantech, Nishinomiya, Japan). The data that were collected during this study and the results obtained from the water analysis were consolidated to generate a map for fecal contamination and water quality parameters using Quantum GIS (QGIS) software 3.32.3. The water quality guidelines for Class C freshwater primary parameters were also used to compare the results, whether they passed or failed the requirements.

3. Results and Discussion

3.1. Flow Velocity

Water quality (WQ) processes can be significantly influenced by flow velocity conditions. Specifically, higher flow velocities are associated with increased reaeration rates in a positive correlation [21]. The flow velocity of each station is shown in Table 3, including their depth and width. The values obtained for flow velocity ranged from 0.22 m/s at Station 9 to 1.27 m/s at Station 7. This was attributed to factors such as steep gradient in the area and narrow channel section in Station 7, which is 8.40 m. On the other hand, Station 9 shows the lowest flow velocity of 0.22 m/s among the recorded stations. This is due to its wider channel section of 12.1 m with a moderate gradient in the area. Additionally, its depth of 0.13 m stands as the lowest depth among all the stations.

Table 3. Velocity, depth, and width of the sampling stations in the Silway River.

Station Number	Velocity (m/s)	Depth (m)	Width (m)
1	0.69	0.55	23.5
2	0.99	0.48	22.9
3	0.92	0.27	6.4
4	0.87	0.54	28.3
5	1.07	0.30	16.5
6	1.17	0.52	24.6
7	1.27	0.35	8.4
8	0.82	0.62	16.9
9	0.22	0.13	12.1
10	0.47	0.35	8.0

The variations in flow velocity along the Silway River can be attributed to several factors, including the river's topography, channel characteristics, and the volume of water being transported. The speed at which water moves through a river may influence several factors and parameters such as dissolved oxygen, suspended solids, temperature, nutrient transport, and pollutants dilution and removal that contribute to its overall quality [22].

3.2. Physicochemical Water Quality Parameters and Chicken Fecal Contamination

The gel image results of the first and second nested PCR of Silway River water samples from Station 1 to Station 10 and a negative control are shown in Figures 6 and 7. Figure 6 shows the agarose gel electrophoresis image result of the first nested PCR of water samples from Silway River stations. On the other hand, Figure 7 shows the agarose gel electrophoresis image result of the second nested PCR of water samples from Silway River stations.

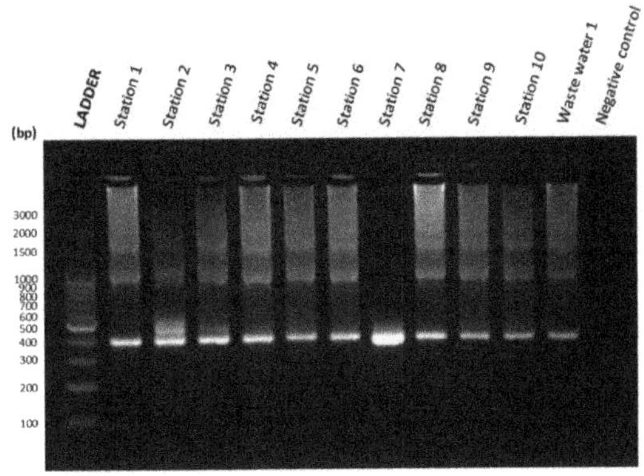

Figure 6. Gel image of first nested PCR of Silway River water samples.

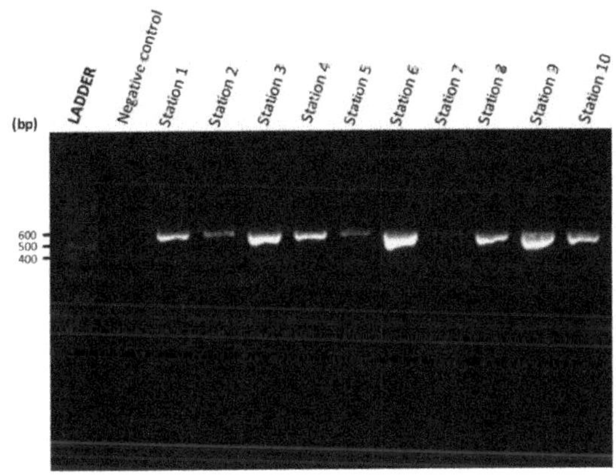

Figure 7. Gel image of second nested PCR of Silway River water samples.

The water quality parameters such as DO, TSS, temperature, pH, turbidity, and PCR and qPCR tests shown in Table 4 were based on the results from on-site fieldwork and the laboratory analysis conducted by the Philippine Genome Center Mindanao Omics Laboratory and Mindanao State University GSC Water Laboratory. The comparison between the results of water quality parameters analysis from ten sampling stations and the standards outlined in the DENR Administrative Order series of 2016-08 are also shown in Table 4.

Table 4. Physicochemical water quality parameters and PCR and qPCR test results for chicken fecal contamination.

Station Number	DO (mg/L)	TSS (mg/L)	Temperature (°C)	pH	Turbidity (FNU)	Fecal Coliform Ckmito PCR Test	Fecal Coliform ND5-CD qPCR Test (Ct Value)
DAO Standard	>5 mg/L	<80 mg/L	25–31	6.5–9.0	-	-	-
1	7.55	181	26.11	8.48	67.947	Positive	29.485
2	8.01	118	25.67	8.20	73.029	Positive	33.850
3	6.82	81	28.06	7.97	68.375	Positive	26.785
4	No data	68	27.52	8.31	29.487	Positive	30.560
5	7.97	186	27.53	8.60	93.532	Positive	33.060

Table 4. *Cont.*

Station Number	DO (mg/L)	TSS (mg/L)	Temperature (°C)	pH	Turbidity (FNU)	Fecal Coliform Ckmito PCR Test	Fecal Coliform ND5-CD qPCR Test (Ct Value)
6	7.68	197	28.72	8.54	123.575	Positive	27.790
7	7.57	631	29.39	8.53	313.084	Positive	35.230
8	7.39	53	31.19	8.30	10.599	Positive	24.590
9	7.66	60	29.06	8.15	98.666	Positive	20.595
10	7.84	1180	26.06	8.35	448.161	Positive	27.665

3.2.1. Dissolved Oxygen (DO)

The measurement of DO is crucial for assessing the metabolic pulses of rivers and is a fundamental indicator of water quality [22]. Figure 8 shows the map for the results of DO concentration along the Silway River.

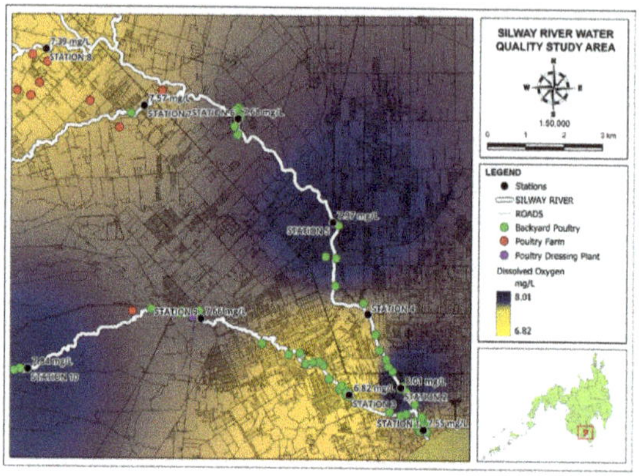

Figure 8. Map of dissolved oxygen results for Silway River stations.

In this study, Stations 2, 5, and 10 have relatively high DO levels with 8.01 mg/L, 7.97 mg/L, and 7.84 mg/L, respectively, while Station 3 has the lowest recorded DO level at 6.82 mg/L. All of the stations contain a DO concentration above the minimum level required, which means that most fish and other living organisms will be able to survive in the water. In the succeeding section, DO will be correlated with chicken fecal contamination.

3.2.2. Total Suspended Solids (TSSs)

From the results shown in Table 4, the majority of the stations have TSS concentrations above the recommended limit of 80 mg/L for Class C freshwater. In Figure 9, Stations 7 and 10 have particularly high TSS levels, with 631 mg/L and 1180 mg/L, respectively. The high concentration of TSSs in the majority of the stations was due to their topography, where the area is characterized by hills and runoff. This is very common, like in Station 10, which is located at the bottom of a hill. On the other hand, only Stations 4, 8, and 9 pass the maximum threshold of 80 mg/L with 68 mg/L, 53 mg/L, and 60 mg/L, respectively. Besides the topographic conditions, flow velocity can also influence the transport and distribution of suspended solids in a river. Historically, in the Silway Water Quality Management Report 2020, 9 out of 25 monitoring stations failed the standard allowable TSS level. This poses a threat to environment safety and health hazards to nearby communities. The accumulation of sediment deposits at river channels not only harbors toxic pollutants but also diminishes the flood-carrying capacity [23].

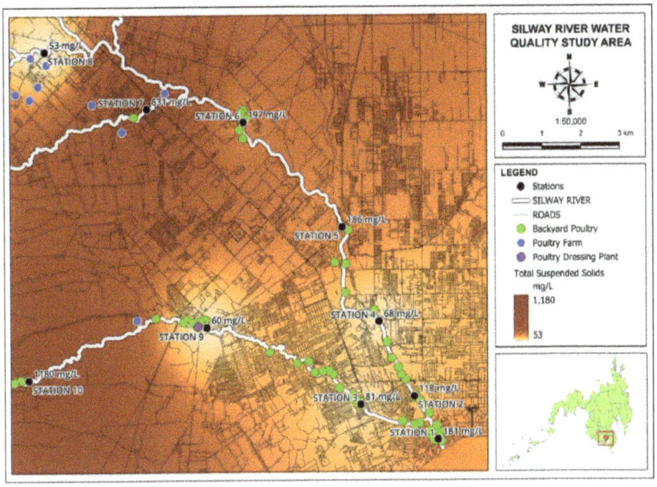

Figure 9. Map of total suspended solids in Silway River stations.

3.2.3. Temperature

The temperature of the water plays a crucial role in shaping the habitat availability for various aquatic species, thereby affecting the survival of some aquatic organisms. It has a significant impact on water quality and various environmental factors, as it governs the diversity of aquatic life, regulates the maximum concentration of dissolved oxygen, and influences the occurrence of chemical and biological reactions [8]. The readings for temperature in the Silway River reveal that the majority of the stations along the river fall within the acceptable temperature range. This indicates that the Silway River generally meets the water quality guidelines for primary parameters at these locations.

Stations 1 and 2 recorded temperatures of 26.11 °C and 25.67 °C, respectively; these values barely fall within the minimum limit. Station 8, by contrast, recorded a temperature of 31.19 °C, which is at the upper limit of the acceptable range. Figure 10 shows a map of the results of temperature.

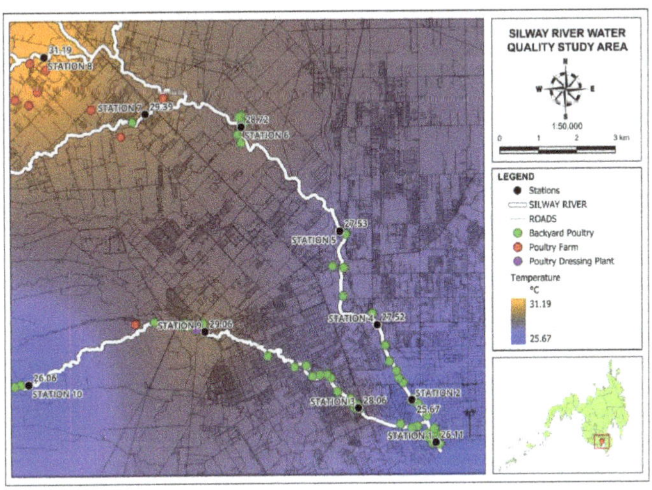

Figure 10. Map of temperature records in Silway River stations.

3.2.4. pH

Low pH levels can lead to the release of toxic elements and compounds, including heavy metals, which can be absorbed by aquatic organisms. In this study, the pH levels recorded at the sampling stations in the given data were generally within the acceptable range, ranging from 7.97 in Station 3 to 8.60 in Station 5. These values indicate that the water in the Silway River is predominantly alkaline in nature. Fluctuations in pH can have detrimental effects on certain species in the river, while others may be more sensitive to such changes. Therefore, monitoring the pH of an aquatic system is essential for assessing the quality of water and pollution levels [8]. Figure 11 shows a map of the results of pH level.

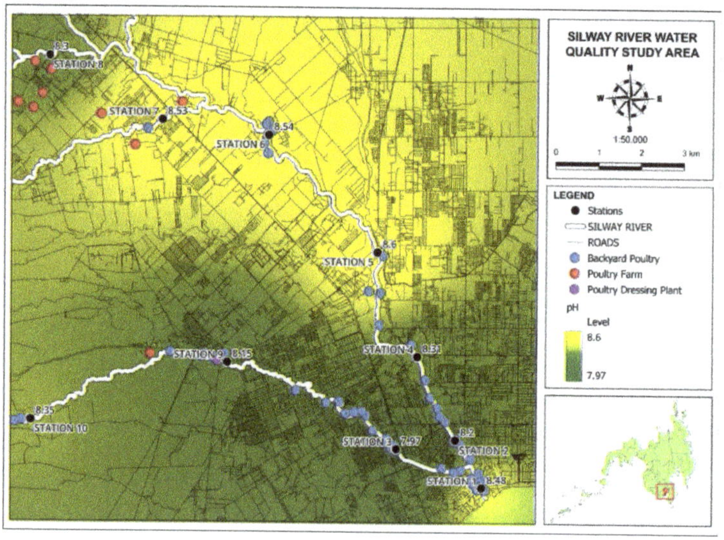

Figure 11. Map of pH levels in Silway River stations.

3.2.5. Turbidity

As shown in Table 4, Station 10 records the highest turbidity value of 448.16 FNU, indicating an extremely high concentration of suspended particles and potentially very cloudy or muddy water. On the other hand, Stations 4 and 8 exhibit lower turbidity readings, with Station 8 having the lowest recorded value of turbidity at 10.60 FNU, suggesting clearer water with fewer suspended particles. Figure 12 shows a map of the turbidity of sampling stations in the Silway River.

Station 10 is an upstream hilly station with mainly pineapple plantations. Pineapple is a famous commodity in the area produced by contract farmers and in the area of a large international company. Research has shown that the rapid deterioration of soil biodiversity on pineapple plantations due to erosion, over-cultivation, and pesticide application eventually leave the land completely infertile [24,25]. Elevated turbidity levels can result in increased heat absorption by the water, leading to higher water temperatures. Additionally, the reduced light penetration caused by high turbidity can diminish photosynthetic activities of aquatic plants and organisms, consequently decreasing the level of dissolved oxygen in the water [26]. Station 7 had the second highest turbidity and had the second highest temperature during the time of data gathering. Station 3 had the lowest DO, and the turbidity level exceeded the minimum allowable standard.

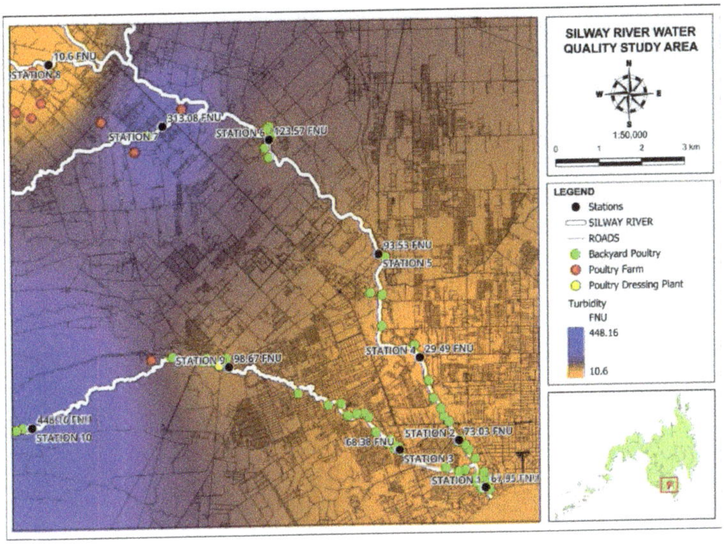

Figure 12. Map of turbidity values in Silway River stations.

3.2.6. Chicken Fecal Contamination

The presence of fecal coliform in water has an adverse effect on the ecosystem and poses a substantial risk to human health [9]. In this study, two poultry-associated gene markers, mtDNA Ckmito and ND5-CD, were used as indicators for microbial source tracking of chicken fecal contamination. The mtDNA marker Ckmito was utilized for MST by PCR to detect chicken fecal contamination. The ND5 gene marker was used for the qPCR assay to quantify chicken fecal contamination. As shown in Table 4, all stations of the Silway River tested positive for chicken fecal contamination in the PCR tests, and all of the samples showed a reaction during the qPCR test. Table 5 further shows the number of poultry farms, backyard poultry, and dressing plants near the station.

Table 5. Poultry farms, backyard poultry, and dressing plants near the stations in the Silway River.

Station Number	Number of Poultry Farm [a], Backyard Poultry [b], and Dressing Plants [c] Within a 1000 m Radius of the Station	Approx. Distance of Structures from the River (m)
1	12 [b]	10–100
2	7 [b]	10–80
3	7 [b]	10–50
4	4 [b]	10–40
5	3 [b]	10–30
6	6 [b]	10–60
7	3 [a], 1 [b]	50–150
8	5 [a]	200–300
9	1 [a], 6 [b], 1 [c]	10–150
10	2 [a], 2 [b]	10–100

[a] Poultry Farm. [b] Backyard Poultry. [c] Dressing Plants.

The Ct (Cycle threshold) value is the number of amplification cycles (utilizing RT-PCR) necessary for the fluorescence of a PCR product to intersect a threshold level, exceeding the background signal [27]. For Ct values, the higher the number, the lower the target gene concentration, while lower Ct values indicate a higher concentration of the target gene. Figure 13 shows a map of the results of chicken fecal contamination in the Silway River.

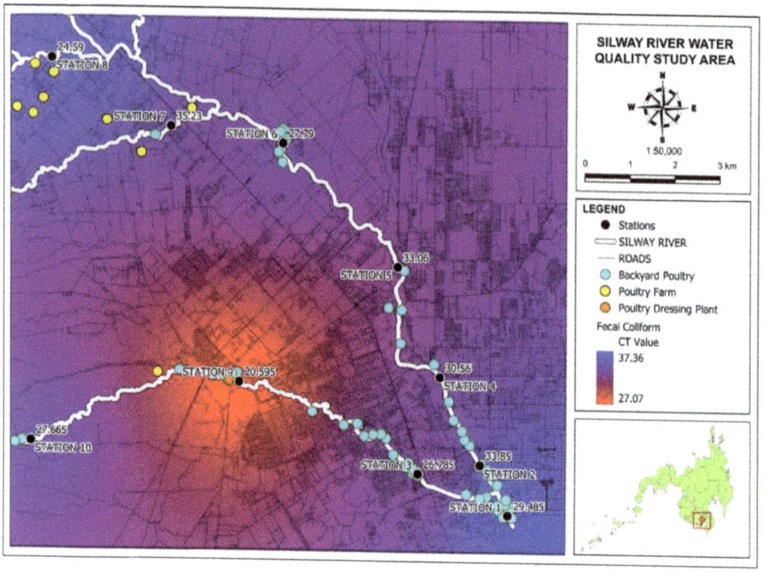

Figure 13. Map of chicken fecal contamination in Silway River stations.

Station 7 has the highest Ct value at 35.23, indicating a low concentration of the marker ND5, while Station 9 and 8 have the lowest Ct values of 20.56 and 24.25, respectively, indicating significantly high gene concentration. The results from PCR and qPCR tests were expected due to the nearby communities and facilities in the Silway River that engaged in poultry raising, especially in Station 9. Backyard poultry raising is very common in various neighborhoods, especially in the province.

The MST method was used in this study to detect and quantify chicken host-specific fecal bacteria using a combination of PCR and qPCR. Based on the results of this study, chicken feces were indeed present and are one of the various sources of fecal contamination that contributes to water pollution in the Silway River. MST is recommended for use especially in studies involving identification and quantification of a specific fecal bacteria not only in water bodies but also in soils. The advantage of MST is that this method has been proven by previous studies to accurately and precisely identify various host-specific sources of fecal contamination in environmental water samples [13]. The tests involved in this method are generally more sensitive and specific than traditional fecal coliform tests.

One of the disadvantages of the MST method is the cost of the tests, which are more expensive to set up, but this can be cost-effective in the long term due to their accuracy and specificity. Table 6 shows a comparison between the cost incurred for the tests in this study and the cost of a traditional fecal coliform detection test.

The cost for a traditional fecal coliform test is based on Mindanao State University—GSC Water Analytical Laboratory and Research Center, while the cost for the tests using the MST method is based on Philippine Genome Center Mindanao Omics Laboratory services. The primers and probe were synthesized by Noveaulab and cost about PHP 29,191.50 as a whole. These primers can analyze at least 700 samples, while probes can analyze at least 250 samples.

Table 6. Cost comparison between MST and a traditional fecal coliform test.

Method	Unit	Price per Unit/Test
Traditional Test		
Detection of Fecal Coliform	samples	PHP 350.00
MST (detection only)		
eDNA Extraction (Water)	samples	PHP 976.60
Gel Electrophoresis	runs	PHP 158.92
PCR Amplification	amplicon	PHP 244.25
PCR Amplification Optimization	amplicon	PHP 244.25
Total		PHP 1624.12
MST (detection and quantification)		
eDNA Extraction (Water)	samples	PHP 976.60
Gel Electrophoresis	runs	PHP 158.92
PCR Amplification	amplicon	PHP 244.25
PCR Amplification Optimization	amplicon	PHP 244.25
qPCR	NC + PC	PHP 112.35
qPCR Optimization	gDNA + NC + PC + NEC	PHP 112.35
Total		PHP 3868.82

3.3. *Relationship Among Flow Velocity, Physicochemical Water Quality Parameters, and Chicken Fecal Contamination*

3.3.1. Chicken Fecal Contamination and Physicochemical Water Quality Parameter Relationship

The rate of coliform bacteria decay tends to rise with higher water temperatures, increased pH levels, nutrient scarcity, and greater levels of predation and parasitism. On the other hand, the survival of coliform bacteria can be extended, or they can even thrive under specific environmental conditions, including appropriate pH, suitable temperature, abundant nutrients, and concentrated suspended particles [28,29] In this study, the levels of temperature and pH fell within the recommended range, while the majority of the results of TSSs failed the requirements. As mentioned from previous studies, these levels of water quality parameters such as temperature and pH and high concentrations of TSSs can be considered favorable conditions for fecal coliforms to grow and survive. Due to this, coliforms coming from chicken feces can stay and survive longer in these environments and continue to multiply. Fecal contamination is significantly high and has always been a problem in the Silway River.

3.3.2. Multiple Regression Analysis of Chicken Fecal Contamination, DO, and Flow Velocity

An inverse relationship was observed between chicken fecal contamination and both DO and flow velocity when examining the data of the Silway River (Table 7).

An R^2 value of 0.98 indicates that higher DO has a higher Ct value, which means less chicken fecal contamination (Figure 14a). Low DO levels could indicate a higher presence of bacteria using it for respiration and an increased amount of decaying organic matter [30]. Meanwhile, an R^2 value of 0.55 and a p-value of 0.01 indicate a significant effect of flow velocity on chicken fecal contamination (Figure 14b). With the availability of a current meter and DO meter, fecal coliform Ct value could be estimated using the equation $Y = -6.920926245 + 3.571535586 \times DO + 10.46724234 \times Velocity$. The equation developed is specific to the Silway River section studied on the given date, reflecting the conditions observed, and has not been validated or applied to other systems.

Table 7. Multiple linear regression analysis results.

Regression Statistics					
Multiple R	0.799465765				
R Square	0.63914551				
Adjusted R Square	0.536044227				
Standard Error	3.04281321				
Observations	10				
ANOVA					
	df	SS	MS	F	Significance F
Regression	2	114.7932244	57.3966122	6.199200363	0.028226849
Residual	7	64.8109856	9.258712228		
Total	9	179.60421			
	Coefficients		Standard Error	t Stat	*p*-value
Intercept	−6.920926245		20.75159251	−0.333513018	0.748514646
DO (mg/L)	3.571535586		2.709846892	1.317984273	0.229000056
Velocity	10.46724234		3.173606926	3.298216378	0.013152948

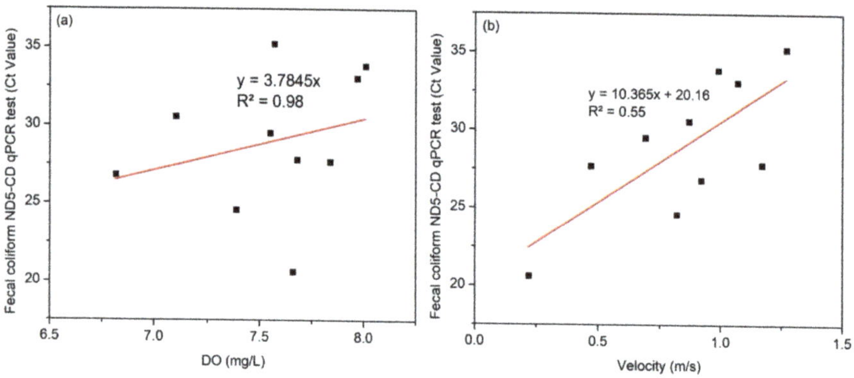

Figure 14. Relationship between fecal coliform Ct value and (**a**) DO (mg/L) and (**b**) velocity (m/s).

The concentration of chicken fecal contamination tends to increase as the flow velocity decreases. This trend was particularly evident in Station 9, which has the lowest Ct value of 20.595, indicating a significantly high concentration of fecal contamination, while flow velocity was lowest at 0.22 m/s. In addition, Station 7 has the highest flow velocity of 1.27 m/s while fecal contamination concentration was lowest with a Ct value of 35.230. This chicken fecal contamination and flow velocity relationship suggests that lower flow velocity may result in the accumulation of fecal contaminants, leading to higher concentrations. Lower flow velocity, particularly in stagnant water conditions, is more favorable for bacterial growth [31] and can promote the settling of fecal contaminants and particulate-bound pathogens in the riverbed. Historical data of the Silway Water Quality Management Report in the year 2020 showed that 25 out of 25 monitoring stations failed the standard allowable fecal coliform level of 200 MPN/100 mL. This means that the fecal coliform contamination problem has been an ongoing, persistent problem that must be addressed.

3.3.3. Turbidity and TSS Relationship

There is a positive linear relationship between turbidity and TSSs. An R^2 value of 0.961 and a p-value of <0.0001 indicates a highly significant effect in the relationship between turbidity and TSSs. As turbidity levels increase, TSS levels also increase. The equation y = 2.3061x was used as the turbidity–TSS conversion equation for all the turbidity data collected on site.

This trend is evident in all stations of the Silway River. Station 10 has the record of the highest values in both turbidity and total suspended solids concentration at 448.16 FNU and 1180 mg/L, respectively, and Station 7 is the second highest with a turbidity value of 313.08 FNU and total suspended solids concentration of 631 mg/L. By contrast, Station 8 has the lowest turbidity and total suspended solids concentration value of 10.60 FNU and 53 mg/L, respectively. This result was expected due to the presence of suspended solids such as organic matter and particles of sediment that cause light to scatter and be absorbed in the water, which can result in increased turbidity. In this study, the positive linear relationship between turbidity and TSSs implies that turbidity can be used as a proxy for estimating TSS levels.

4. Conclusions

Molecular analyses, such as qPCR using host-specific markers (ND5-CD, Ckmito) and MST methods, were used along with river water parameter analysis to investigate fecal contamination in the Silway River. The water samples collected from all stations tested positive for the Ckmito marker during the first and nested PCR tests, and all water samples exhibited a reaction crossing the threshold during the qPCR test. The results of this study suggest that nearby households engaged in backyard poultry raising and poultry farms are contributing to the chicken fecal contamination in the Silway River. In addition, the MST method combining Ckmito and ND5 markers proved effective in detecting and quantifying host-specific fecal bacteria in water bodies. However, the MST method is more expensive to set up, although it is considered cost-effective in the long term due to its accuracy and specificity.

Physicochemical parameters such as dissolved oxygen, temperature, and pH were all within the standard limit in all stations except for total suspended solids, where the majority of the sampling stations fell above the maximum limit. Although turbidity was not mentioned in the DENR standard water quality guidelines, the results in this study showed significantly high levels of turbidity. The DO levels in all stations were generally above the minimum required for aquatic life, which is >5 mg/L. The temperature fell within the acceptable range at all stations, but Station 8 was at the upper limit at 31.19 °C. pH levels were on the basic side, ranging from 7.97 to 8.60, indicating mainly alkaline water. TSS and turbidity levels were exceptionally high in Stations 7 and 10. These elevated TSS and turbidity levels are largely attributed to the topography of the study area, characterized by hills and runoff. This situation was evident in Station 10, which is located at the foot of a hill, suggesting cloudy or muddy water.

Relationships among flow velocity, water quality parameters, and chicken fecal contamination were also studied. Chicken fecal contamination and water quality parameters relationship suggests that normal levels of temperature and pH, along with high concentrations of TSSs, create a favorable condition for fecal coliforms to thrive. Chicken fecal contamination and flow velocity showed a significant inverse relationship, with an R^2 value of 0.55 and p-value of 0.01, suggesting that lower flow velocity may result in the accumulation of fecal contamination, leading to higher concentrations and contributing to persistent fecal contamination in the Silway River. Similarly, DO had an inverse relationship with chicken fecal contamination since lower Ct values mean higher contamination. A strong coefficient of determination was found between DO and chicken fecal contamination Ct value, where R^2 was 0.98 with a p-value of 0.2, suggesting that low DO means higher fecal contamination. Lastly, turbidity and TSS exhibited a strong positive linear relationship, with an R^2 value of 0.961 and a p-value of <0.0001, such that higher turbidity levels were associated with elevated TSS levels. Turbidity can be used as a proxy for estimating TSS levels.

While direct relationships between fecal contamination and river water parameters were obtained, the molecular tools used were significant to pinpoint a specific fecal contributor, which is poultry. The results are essential for targeted remediation strategies. High-order riverine systems are influenced by multiple tributaries and diverse inputs, making it challenging to associate fecal contamination with individual upstream activities. MST and qPCR methods address this complexity by using source-specific markers, even in matrices with low fecal loads, where conventional approaches fail.

The results of this study were, however, derived from water samples collected during a single day under specific meteorological conditions. Environmental variables, such as prolonged drought or typhoon events, could substantially alter the outcomes. Drought conditions may reduce surface runoff, which could lead to undetectable levels of chicken markers. By contrast, typhoon-induced runoff could significantly elevate fecal contamination levels due to a high influx of water from more pollution sources. Therefore, these findings are limited by time and location and should be interpreted in the context of the environmental conditions during the sampling.

In conclusion, this study confirms the presence of chicken fecal contamination in the Silway River and emphasizes the significance of the application of the MST method for detecting and quantifying chicken-specific fecal bacteria in water bodies. With the growing accessibility of qPCR technologies, molecular analyses are pivotal for both addressing chronic water quality impairments and preventing contamination. This is particularly valuable, as communities face escalating challenges from both point and non-point pollution sources. Nearby households engaged in backyard poultry raising and poultry farms are one of the main sources of chicken fecal contamination in the Silway River. Lastly, flow velocity and water quality parameters play a significant role in the survival of certain fecal bacteria in environmental waters. The MST and eDNA approaches could be utilized in future research to assess the transport and persistence of pathogens in river systems, alongside hydrodynamic flow models, to offer a comprehensive understanding of contamination dynamics over time and space.

Author Contributions: Conceptualization, L.M.O. and J.C.C., methodology, L.M.O. and J.C.C., formal analysis, L.M.O., J.C.C., R.L., M.S. and A.B., investigation, L.M.O. and J.C.C., resources, L.M.O., J.C.C., K.Y. and S.H., writing—original draft preparation, L.M.O., writing—review and editing, L.M.O., J.C.C., R.L., M.S., A.B., K.Y. and S.H., visualization, L.M.O. and J.C.C., supervision, J.C.C. All authors have read and agreed to the published version of the manuscript.

Funding: This research was funded by the Department of Science and Technology—Engineering Research and Development for Technology (DOST-ERDT) scholarship of the first author.

Institutional Review Board Statement: Not applicable.

Informed Consent Statement: Not applicable.

Data Availability Statement: The original contributions presented in the study are included in the article, further inquiries can be directed to the corresponding author.

Acknowledgments: The authors would like to thank everyone who helped during the administration of the reconnaissance survey and actual fieldwork, Danilo Opog and Alona Opog for preliminary site inspection, John Miro Bataan for their support, and Christian Jake Lubaton, Aldbrian Christon Ang, Adrian Camain, and John Renelle Carandang, who assisted during the field measurements and collection of water samples from the river.

Conflicts of Interest: The authors declare no conflicts of interest.

References

1. Dili, R.; Kalaw, R.; Miguel, A.; Ting, G. Analysis of Environmental Impact and Waste Management of Egg Poultry Industry in the Philippines: A Case of San Jose, Batangas. *J. Sustain. Environ. Manag.* **2022**, *1*, 188–196. [CrossRef]
2. PSA. *Chicken Situation Report April to June 2022*; Philippine Statistics Authority: Quezon City, Philippines, 2022.

3. Medina, B.O.; De Torres, E.M.; Dela Peña, A.R. Waste Disposal Practices of Backyard Poultry Owners in San Jose, Batangas, Philippines. *Int. J. Adv. Res. Publ.* **2019**, *3*, 194–198.
4. Gržinić, G.; Piotrowicz-Cieślak, A.; Klimkowicz-Pawlas, A.; Górny, R.L.; Olkowska, E.; Potrykus, M.; Tankiewicz, M.; Krupka, M.; Siebielec, G.; Wolska, L. Intensive poultry farming: A review of the impact on the environment and human health. *Sci. Total Environ.* **2023**, *858*, 160014. [CrossRef]
5. Zhuang, F.F.; Li, H.; Zhou, X.Y.; Zhu, Y.G.; Su, J.Q. Quantitative detection of fecal contamination with domestic poultry feces in environments in China. *AMB Express* **2017**, *7*, 80. [CrossRef] [PubMed]
6. WHO. *Guidelines for Drinking-Water Quality (Incorporating 1st and 2nd Addenda)*; WHO: Geneva, Switzerland, 2008.
7. Smith, V.H.; Schindler, D.W. Eutrophication science: Where do we go from here? *Trends Ecol. Evol.* **2009**, *24*, 201–207. [CrossRef] [PubMed]
8. Escatron, M.E.; Villamor, V.F.; Eviota, M.P.; Escatron, R.A. Water quality assessment of Surigao River, Surigao City, Philippines. *J. Entomol. Zool. Stud.* **2022**, *10*, 1–10. [CrossRef]
9. Paruch, L.; Paruch, A.M. An Overview of Microbial Source Tracking Using Host-Specific Genetic Markers to Identify Origins of Fecal Contamination in DifferentWater Environments. *Water* **2022**, *14*, 1809. [CrossRef]
10. Stoeckel, D.M.; Harwood, V.J. Performance, Design, and Analysis in Microbial Source Tracking Studies. *Appl. Environ. Microbiol.* **2007**, *73*, 2405–2415. [CrossRef]
11. Bautista, J.; Manubag, J.; Sumaya, N.; Martinez, J.; Tabugo, S. Environmental DNA (eDNA) metabarcoding and fish visual census reveals the first record of Doboatherina magnidentata in the Philippines. *Biodiversitas J. Biol. Divers.* **2023**, *24*, 3063–3072. [CrossRef]
12. Dela Peña, L.O.; Labrador, K.L.; Nacario, M.G.; Bolo, N.R.; Rivera, W.L. Microbial source tracking of fecal contamination in Laguna Lake, Philippines using the library-dependent method, rep-PCR. *J. Water Health* **2021**, *19*, 762–774. [CrossRef]
13. Feng, S.; Bootsma, M.; McLellan, S.L. Human-Associated Lachnospiraceae Genetic Markers Improve Detection of Fecal Pollution Sources in Urban Waters. *Appl. Environ. Microbiol.* **2018**, *84*, 309–318. [CrossRef] [PubMed]
14. Hachad, M.; Burnet, J.P.; Sylvestre, E.; Duy, S.V.; Villemur, R.; Sauvé, S.; Prévost, M.; Qiu, J.Y.; Pang, X.; Dorner, S. β-D-glucuronidase activity triggered monitoring of fecal contamination using microbial and chemical source tracking markers at drinking water intakes. *Water Res.* **2024**, *254*, 121374. [CrossRef] [PubMed]
15. Golder, A.R. Quantification of Poultry and Human Fecal Contamination in the Tidal Creeks of the Virginia Eastern Shore Using Multifaceted eDNA Method. Ph.D. Thesis, William & Mary, Williamsburg, VA, USA, 2023.
16. Lim, J.M.U. Water Pollution Prevention and Control Program: The Polomolok Experience. Department of Environment and Natural Resources (DENR) Philippines, Quezon City 2010. Available online: https://www.scribd.com/document/535058811/Water-Pollution-and-Prevention-0 (accessed on 17 September 2023).
17. Asian Development Bank (ADB). 2022. Available online: https://bimp-eaga.asia/sites/default/files/publications/Reduced%20PRF18-IP%20GCAP%20General%20Santos%20City_WEB_28Nov.pdf (accessed on 1 December 2024).
18. Suarez, M.N.S.; Zoleta, J.M.R. Water Quality Assessment of Sarangani Bay: Basis for Sustainable Coastal Management. *Am. J. Environ. Clim.* **2024**, *3*, 34–44. [CrossRef]
19. Barro, R.I.; Puno, G.R. Morphometric Analysis of Silway River Basin in Southern Mindanao, Philippines for Flood Risk Management a Supplementary of Flood Modeling. Asian Assoc. Remote Sens. 2016. Available online: https://www.scribd.com/document/671218201/Ab-0527 (accessed on 17 September 2023).
20. Liang, H.; Yu, Z.; Ndayisenga, F.; Liu, R.; Zhang, Y.; Zhang, H.; Wu, G. A combination of mitochondrial DNA markers Ckmito and ND5-CD is recommended as the most reliable indicator for microbial source tracking to identify faecal pollution from poultry in China. *Ecol. Indic.* **2020**, *115*, 106334. [CrossRef]
21. Assar, W. Quantitative study for the effect of water velocity on water quality change. *Eur. J. Sci. Technol.* **2023**, *47*, 37–41. [CrossRef]
22. Casila, J.C.; Azhikodan, G.; Yokoyama, K. Quantifying water quality and flow in multi-branched urban estuaries for a rainfall event with mass balance method. *Water Sci. Eng.* **2020**, *13*, 317–328. [CrossRef]
23. Zhi, W.; Feng, D.; Tsai, W.P.; Sterle, G.; Harpold, A.; Shen, C.; Li, L. From hydrometeorology to river water quality: Can a deep learning 3 model predict dissolved oxygen at the continental scale? *Environ. Sci. Technol.* **2021**, *55*, 2357–2368. [CrossRef] [PubMed]
24. Casila, J.C.; Andres, H.L.; Haddout, S.; Yokoyama, K. Sediment Transport Modeling in the Pasig River, Philippines Post Taal Volcano Eruption. *Geosciences* **2024**, *14*, 45. [CrossRef]
25. Brown, J.; Flint, T.; Lamay, J. The Politics of Pineapple: Examining the Inequitable Impacts of Southern Costa Rica's Pineapple Industry. *J. Public Int. Aff.* **2020**, 33. Available online: https://jpia.princeton.edu/news/politics-pineapple-examining-inequitable-impacts-southern-costa-ricas-pineapple-industry (accessed on 27 November 2024).
26. Richardson, R.B.; Kellon, D.; Leon, R.G.; Arvai, J. Using choice experiments to understand household tradeoffs regarding pineapple production and environmental management in Costa Rica. *J. Environ. Manag.* **2013**, *127*, 308–316. [CrossRef] [PubMed]
27. EPA. United States Environmental Protection Agency. 2021. Available online: https://www.epa.gov/awma/factsheets-water-quality-parameters (accessed on 10 May 2023).

28. Public Health Ontario. Ontario Agency for Health Protection and Promotion. Queen's Printer for Ontario. 2020. Available online: https://www.publichealthontario.ca/-/media/documents/ncov/main/2020/09/cycle-threshold-values-sars-cov2-pcr.pdf?la=en (accessed on 21 October 2023).
29. Hong, H.; Qiu, J.; Liang, Y. Environmental factors influencing the distribution of total and fecal coliform bacteria in six water storage reservoirs in the Pearl River Delta Region, China. *J Environ. Sci.* **2010**, *22*, 663–668. [CrossRef] [PubMed]
30. Pearson, H.W.; Mara, D.D.; Mills, S.W.; Smallman, D.J. Physico-Chemical Parameters Influencing Faecal Bacterial Survival in Waste Stabilization Ponds. *Water Sci. Technol.* **1987**, *19*, 145–152. [CrossRef]
31. Casila, J.C.; Ella, V.; Yokoyama, K. Effect of neap-spring hydrodynamics on salt and suspended sediment transport in multi-branched urban estuaries. *Sci. Total Environ.* **2019**, *694*, 133653. [CrossRef] [PubMed]

Journal of *Xenobiotics*

MDPI

Article

Impact of Short-Chain Perfluoropropylene Oxide Acids on Biochemical and Behavioural Parameters in *Eisenia fetida* (Savigny, 1826)

Davide Rotondo [1], Davide Gualandris [1], Candida Lorusso [1], Albert Braeuning [2], Antonio Calisi [1] and Francesco Dondero [1,*]

1 Department of Science and Technological Innovation (DISIT), University of Eastern Piedmont, Viale Michel 11, 15121 Alessandria, Italy; davide.rotondo@uniupo.it (D.R.); davide.gualandris@uniupo.it (D.G.); candida.lorusso@uniupo.it (C.L.); antonio.calisi@uniupo.it (A.C.)

2 Department of Food Safety, German Federal Institute for Risk Assessment, Max-Dohrn-Str. 8-10, 10589 Berlin, Germany; albert.braeuning@bfr.bund.de

* Correspondence: francesco.dondero@uniupo.it

Abstract: Per- and polyfluoroalkyl substances (PFAS) are a class of persistent organic pollutants that pose a growing threat to environmental and human health. Soil acts as a long-term reservoir for PFAS, potentially impacting soil biodiversity and ecosystem function. Earthworms, as keystone species in soil ecosystems, are particularly vulnerable to PFAS exposure. In this study, we investigated the sublethal effects of three short-chain (C4–C6) next-generation perfluoropropylene oxide acids (PFPOAs) on the earthworm *Eisenia fetida*, using a legacy perfluoroalkyl carboxylic acid (PFCA), perfluorooctanoic acid (PFOA), as a reference. We assessed a suite of biochemical endpoints, including markers for oxidative stress (catalase and superoxide dismutase activity), immunity (phenol oxidase activity), neurotoxicity (acetylcholinesterase activity), and behavioural endpoints (escape test). Results indicate that all tested PFAS, even at sub-micromolar concentrations, elicited significant effects across multiple physiological domains. Interestingly, HFPO-DA demonstrated the most substantial impact across all endpoints tested, indicating broad and significant biochemical and neurotoxic effects. Our findings underscore the potential risks of both legacy and emerging PFAS to soil ecosystems, emphasising the need for further research to understand the long-term consequences of PFAS contamination.

Keywords: acetylcholinesterase; catalase; immune response; phenol oxidase; poly and perfluoroalkyl substances; superoxide dismutase

1. Introduction

Per- and polyfluoroalkyl substances (PFAS) are persistent organic pollutants that have become a global health and environmental concern [1]. These chemicals, used in various industrial and consumer products, are ubiquitous in the environment, detected in air, water, soil, and biota [2,3]. Legacy PFAS, such as perfluorooctanoic acid (PFOA) and perfluorooctanesulfonic acid (PFOS), have been widely used for decades and are now subject to increased regulatory restrictions due to their persistent and bioaccumulative nature. In contrast, emerging PFAS are newer compounds developed to replace legacy PFAS and are often under-researched in terms of their environmental and health impacts. Major sources of PFAS release include industrial processes, firefighting foams, landfills, and wastewater treatment plants [4]. Due to their persistence and mobility, PFAS can accumulate in organisms and biomagnify through food webs [5,6].

Soil acts as a significant reservoir for PFAS [7]. The bioavailability of PFAS in soil depends on various factors, including soil properties, PFAS chain length, and the presence of other contaminants [8–11]. Longer-chain PFAS tend to be more strongly adsorbed to soil particles, while shorter-chain PFAS may leach more easily into groundwater or volatilise into the air [12,13].

PFAS contamination may pose a risk to soil biodiversity and ecosystem function. Earthworms are particularly vulnerable as a keystone species in soil ecosystems. These organisms play crucial roles in nutrient cycling, organic matter decomposition, and soil structure maintenance [14]. The toxic effects of PFAS on earthworms can disrupt these ecological functions and potentially impact higher trophic levels.

While acute toxicity of PFAS in earthworms and mammals, including humans, is relatively low [15,16], recent studies have highlighted their potential for sublethal effects across a wide range of organisms [17]. The International Agency for Research on Cancer (IARC) recently classified PFOA, a legacy perfluoroalkyl carboxylic acid (PFCA), as a Group 1 carcinogen due to its long-term toxicity in humans [18]. This underscores the importance of investigating sublethal mechanisms and chronic exposure effects of both legacy and emerging PFAS. Earthworms are valuable model organisms for assessing PFAS ecotoxicology [5,15,19,20]. They can readily accumulate a range of PFAS, including short-chain congeners, precursors, and fluorotelomers [6,19,21]. Biochemical and behavioural biomarkers in earthworms offer a sensitive tool for monitoring soil health and assessing the sublethal impacts of PFAS contamination [22–25].

In this study, we investigated the sublethal effects of PFAS on the earthworm *Eisenia fetida*. These included three perfluoropropylene oxide acids (PFPOAs) hexafluoropropylene oxide dimer acid (HFPO-DA, also known as GenX), perfluoro(2-methoxy)propanoic acid (PFMOBA), and perfluoro(3-methoxy)propanoic acid (PFMOPrA) along with perfluorooctanoic acid (PFOA) as a reference compound. HFPO-DA is widely used as a substitute for PFOA in the production of polytetrafluoroethylene (PTFE), and during this critical industrial process, PFMOBA and PFMOPrA may form as by-products [26]. Structurally, the three PFPOAs differ from each other by only one carbon in their chain length, enabling a comparative assessment of how specific alterations in molecular configuration influence their environmental behaviour and toxicity. By examining these closely related compounds, we aim to provide a deeper understanding of their modes of action, thereby informing risk assessments and guiding future regulatory decisions.

We exposed earthworms to PFAS through contact for 72 h and assessed a suite of biochemical endpoints (catalase, superoxide dismutase, phenol oxidase, and acetylcholinesterase activities) alongside behavioural responses, each reflecting distinct aspects of earthworm physiology. This research aims to enhance our understanding of the sublethal toxicity of emerging PFAS and contribute to a comprehensive risk assessment of these ubiquitous contaminants.

2. Materials and Methods

2.1. Chemicals

PFAS congeners of the highest available purity were purchased from the following suppliers:

2,2,3,3,4,4,5,5,6,6,7,7,8,8,8-pentadecafluorooctanoic acid (PFOA, CAS No. 335-67-1, 95%) from Merck (Darmstadt, Germany);

2,3,3,3-tetrafluoro-2-(1,1,2,2,3,3,3-heptafluoropropoxy)propanoic acid (HFPO-DA, CAS No. 13252-13-6, 97%) from SynQuest Laboratories, Inc. (Alachua, FL, USA);

2,2,3,3-tetrafluoro-3-(trifluoromethoxy)propanoic acid (PFMOPrA, CAS No. 377-73-1, 98%) from SynQuest Laboratories, Inc. (Alachua, FL, USA); and

2,2,3,3,4,4-hexafluoro-4-(trifluoromethoxy)butanoic acid (PFMOBA, CAS No. 863090-89-5, 97%) from Apollo Scientific (Bredbury, UK).

All other chemicals were of analytical grade.

2.2. Animals and Treatments

Eisenia fetida (Savigny, 1826) earthworms were cultured in a controlled environment (20 ± 2 °C, complete darkness) using a 1:1 mixture of sphagnum peat and organic garden soil, moistened to approximately 50% of its maximum water-holding capacity with deionised water. The pH was maintained at 6.0 ± 1.0 using calcium carbonate ($CaCO_3$). Earthworms were fed weekly with manure from healthy, medication-free horses sourced from a local farm. A homogenous culture of adult earthworms (wet weight after gut content clearance: 0.58 ± 0.08 g, mean ± standard error) was obtained for the experiments. Adult earthworms were allowed to lay cocoons in fresh substrate for 4 weeks and then removed. The resulting juveniles were reared to adulthood (indicated by the appearance of a visible clitellum) for approximately 16 weeks.

Individual adult earthworms were placed in 7 cm plastic Petri dishes lined with a 3M No. 3 filter paper moistened with 1 mL of the respective PFAS solution or vehicle control (0.005% 2-propanol). Exposures were conducted for 72 h according to a modified OECD Test Guideline No. 207 filter paper contact test [27]. Stock solutions of PFOA, HFPO-DA, PFMOBA, and PFMOPrA were prepared in 2-propanol at 0.54 M [28] and then diluted in a 171 mOsmol kg^{-1} physiological solution, pH 7.3 [29], to achieve the desired test concentrations (229 μM, 31 μM, 4.2 μM, 0.6 μM). The concentration of 2-propanol in the control was minimised to match the concentration used in the test solutions.

The PFAS concentration range was determined through preliminary mortality tests conducted in accordance with OECD 207 guidelines, which confirmed that the tested PFAS congeners were non-lethal at concentrations up to the millimolar range. Sublethal concentrations were selected in the micromolar range to reflect levels reported in environments contaminated by AFFF-derived PFAS. The geometric progression, with a factor of e^2 (where e is Euler's number, approximately 2.718, and $e^2 \approx 7.389$), was initiated from environmentally measured levels of total PFAS found in soil at a pilot site studied within the framework of the SCENARIOS project (https://scenarios-project.eu) (accessed on 1 November 2024), ensuring an even distribution across the range while maintaining environmental relevance. Five experimental groups were established for each PFAS, with each group exposed to a different concentration. Each group consisted of 10 replicates (Petri dishes), with one earthworm per dish. Exposures were conducted at 20 °C in the darkness, with humidity controlled to prevent filter paper desiccation. Following the 72h exposure, earthworms were immediately subjected to a behavioural escape test (see Section 2.6) and then stored at −80 ± 1 °C until further analysis.

2.3. Biochemical Determinations

Oxidative Stress Biomarkers

Earthworms were homogenised in an ice-cold buffer (100 mM Tris-HCl, pH 7.6, 0.1% Triton X-100) using a glass–Teflon Potter–Elvehjem homogeniser. Homogenates were centrifuged at 10,000× *g* for 20 min at 4 °C, and the resulting supernatant (SN10) was collected. Catalase (CAT) activity in SN10 was determined using a native in-gel assay [30]. Total protein content in SN10 was determined using the Bradford assay with bovine serum albumin as a standard. Briefly, SN10 samples were diluted to a final protein concentration of 7.5 μg per sample in 0.05 M Tris-HCl, pH 6.8, 50% glycerol, 0.05% bromophenol blue. Samples and bovine erythrocytes catalase (1 μg) were loaded onto precast non-denaturing 12% polyacrylamide gels (Mini-PROTEAN® TGX™, Bio-Rad Laboratories, Hercules, CA,

USA) and electrophoresed at 150 V for 90 min at 4 °C in running buffer, 0.05 M Tris-HCl, pH 8.3, 0.3 M glycine.

After electrophoresis, gels were washed three times for 5 min in ultrapure water (18 MΩ·cm) and then incubated for 10 min in a freshly prepared 0.01% hydrogen peroxide solution diluted from a 30% stock (Merck, Darmstadt, Germany). Gels were washed again twice in ultrapure water and then incubated with gentle agitation in a freshly prepared staining solution containing 2% ferric chloride hexahydrate and 2% potassium ferricyanide trihydrate (Merck, Darmstadt, Germany). Catalase activity appeared as clear bands against a blue-green background. Band intensities were semi-quantified by densitometry using ImageJ v.1.54K software and compared to a standard curve generated with bovine catalase.

Superoxide dismutase (SOD) activity was also determined in SN10 using a competitive assay. This assay, based on the inhibition of tetrazolium salt reduction by superoxide radicals, was performed using a commercial reagent system (SOD Assay Kit, Merck, Darmstadt, Germany) and an SOD standard curve for quantification. Each sample in the 96-well plate contained 7.5 μg of protein in a final volume of 100 μL.

2.4. Phenol Oxidase Activity

Phenol oxidase (PO) activity was determined in the SN10 supernatant. The assay was performed in 96-well microplates using a reaction mixture containing 100 mM Tris-HCl buffer, pH 8.0, 50 mM $CaCl_2$, and 10 mM L-DOPA, as reported by Prochazkova et al. (2019) [31]. Eighty microliters (80 μL) of this reaction mixture were combined with 7.5 μg of protein in 10 μL of each SN10 sample. The microplate was then incubated for 4 h at 25 °C in a dark environment. Following incubation, the absorbance at 492 nm was measured using an Infinite F200 PRO spectrophotometer (Tecan Group Ltd., Männedorf, Switzerland).

2.5. Acetylcholinesterase Activity

Acetylcholinesterase (AChE) activity was determined spectrophotometrically in the SN10 supernatant using a commercially available kit (Acetylcholinesterase Assay Kit, Ikzus, Italy) adapted to a microplate format. Briefly, 7.5 μg of protein in a final volume of 200 μL was added to each well. The assay was based on the conversion of acetylthiocholine to thiocholine by AChE, which subsequently reacts with Ellman's reagent (5,5′-dithiobis-(2-nitrobenzoic acid), DTNB) to produce a yellow product. The reaction kinetics were monitored at 405 nm for 10 min using a microplate reader, and the AChE activity was calculated as nmol/min/mg protein after pathlength correction.

2.6. Escape Test

The escape test was conducted in a circular plastic arena (approximately 40 cm in diameter) containing an inverted 9 cm Petri dish lid placed in the centre. One millilitre (1 mL) of 200 mM NaCl solution was deposited onto the centre of the lid. An individual earthworm was then placed directly into the NaCl droplet, and the time taken for the earthworm to completely leave the droplet was recorded. A maximum escape time of 20 min was set; if an earthworm did not exit within this time frame, the escape time was recorded as 20 min. This escape test protocol was a modification of the method described by Da Cruz Jung et al. (2021) [32]. After each trial, both the arena and Petri dish lid were thoroughly rinsed with tap water, followed by distilled water to remove any residual mucus that could potentially influence subsequent tests.

2.7. Statistical Analysis

Statistical analysis of the data was performed using GraphPad Prism™ 9 (GraphPad Software, San Diego, CA, USA) and Systat 12 (Systat Software Inc., San Jose, CA, USA). Data were first assessed for normality and homoscedasticity. Normally distributed data with equal variances were analysed using one-way analysis of variance (ANOVA), followed by Holm–Sidak's multiple comparison test. In cases of unequal variances, Brown–Forsythe ANOVA with Dunnett's T3 post hoc test was applied. For data that did not meet the assumptions of normality, the non-parametric Kruskal–Wallis test was used, followed by either Dunn's multiple comparison test or an uncorrected Dunn's multiple comparison test.

3. Results

3.1. Antioxidant Enzyme

3.1.1. Catalase Activity

The effects of a 72 h exposure to PFOA and three PFPOA congeners—HFPO-DA, PF-MOBA, and PFMOPrA—on the catalase (CAT) activity of *Eisenia fetida* tissue supernatants are presented in Figure 1.

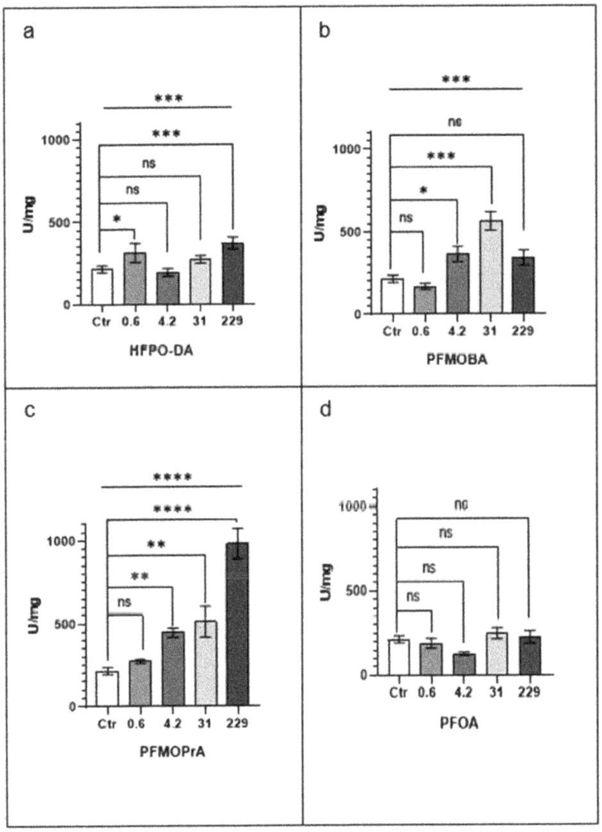

Figure 1. CAT (catalase) activity. Data are presented as the mean activity per mg of protein ± standard error of the mean (SEM). Statistical significance was determined by the Kruskal–Wallis test followed by a post hoc Dunn's multiple comparison test. * $p < 0.05$; ** $p < 0.01$; *** $p < 0.001$; **** $p < 0.0001$; ns, not statistically significant. The x-axis reports PFAS concentrations in μM, and Ctr represents vehicle-exposed earthworms. Lines above the histogram bars indicate statistical significance between PFAS-treated groups and the control. Each panel represents results for a different PFAS compound: (**a**) HFPO-DA, (**b**) PFMOBA, (**c**) PFMOPrA, and (**d**) PFOA.

Overall, exposure to PFPOA congeners resulted in significant increases in catalase activity compared to the control, whereas PFOA did not induce a consistent trend. Among the PFPOAs, PFMOPrA exhibited the most pronounced and concentration-dependent effect.

At 0.6 μM, only HFPO-DA caused a significant increase in catalase activity compared to the control ($p < 0.05$). PFMOBA, PFMOPrA, and PFOA did not show significant changes at this concentration. At 4.2 μM, both PFMOBA and PFMOPrA significantly increased catalase activity ($p < 0.05$ and $p < 0.01$, respectively), while HFPO-DA and PFOA did not exhibit significant effects. PFMOPrA's effect was more substantial than that of PFMOBA at this concentration. At 31 μM, PFMOBA and PFMOPrA continued to show significant enhancements in catalase activity ($p < 0.001$ and $p < 0.01$, respectively). HFPO-DA did not elicit significant changes at this concentration, and PFOA remained without effect. At 229 μM, PFMOPrA induced the most pronounced increase in catalase activity, tripling enzyme functionality compared to the control ($p < 0.0001$). HFPO-DA also showed a significant increase at this concentration ($p < 0.001$), whereas PFMOBA did not produce a significant effect. PFOA did not significantly affect catalase activity at this concentration.

In contrast to the PFPOA congeners, PFOA exposure did not result in significant increases in catalase activity at any concentration. At 4.2 μM, PFOA exposure led to a decrease in catalase activity that approached statistical significance ($p = 0.05$).

Comparatively, PFMOPrA demonstrated the most consistent and robust increase in catalase activity across all concentrations, indicating a clear concentration-dependent effect. PFMOBA also showed significant increases at intermediate concentrations but lacked a consistent dose-response trend. HFPO-DA's effects were significant only at the lowest and highest concentrations, without a clear pattern. The lack of effect by PFOA suggests a distinct mode of action compared to the PFPOA congeners.

These findings indicate that PFPOA congeners, particularly PFMOPrA, may induce oxidative stress responses in earthworms, as evidenced by increased catalase activity, whereas PFOA does not elicit a similar response.

3.1.2. Superoxide Dismutase Activity

Figure 2 illustrates the effects of PFAS exposure on superoxide dismutase (SOD) activity in *Eisenia fetida* tissue extracts.

PFOA treatment resulted in an overall increase in SOD activity, primarily driven by a significant enhancement at the lowest concentration tested (0.6 μM, $p < 0.001$). No significant changes were, in fact, observed at higher concentrations.

HFPO-DA exposure caused a consistent and significant decrease in SOD activity across all concentrations. The most substantial reduction occurred at 31 μM ($p < 0.001$), with significant decreases to a similar extent also noted at 0.6 μM, 4.2 μM, and 229 μM ($p < 0.05$). This suggests that HFPO-DA inhibits SOD activity in a concentration-independent manner.

PFMOBA and PFMOPrA exposures did not produce statistically significant changes in SOD activity at any concentration, indicating a lack of effect on this enzyme.

Comparing the compounds, PFOA and HFPO-DA exhibited opposing effects on SOD activity. PFOA increased SOD activity at low concentrations, while HFPO-DA decreased it across all concentrations. PFMOBA and PFMOPrA did not significantly affect SOD activity, suggesting they do not impact this aspect of the antioxidant defence system.

These results imply that PFOA and HFPO-DA may modulate oxidative stress pathways differently, with PFOA potentially enhancing and HFPO-DA inhibiting the earthworm's ability to mitigate oxidative stress through SOD activity.

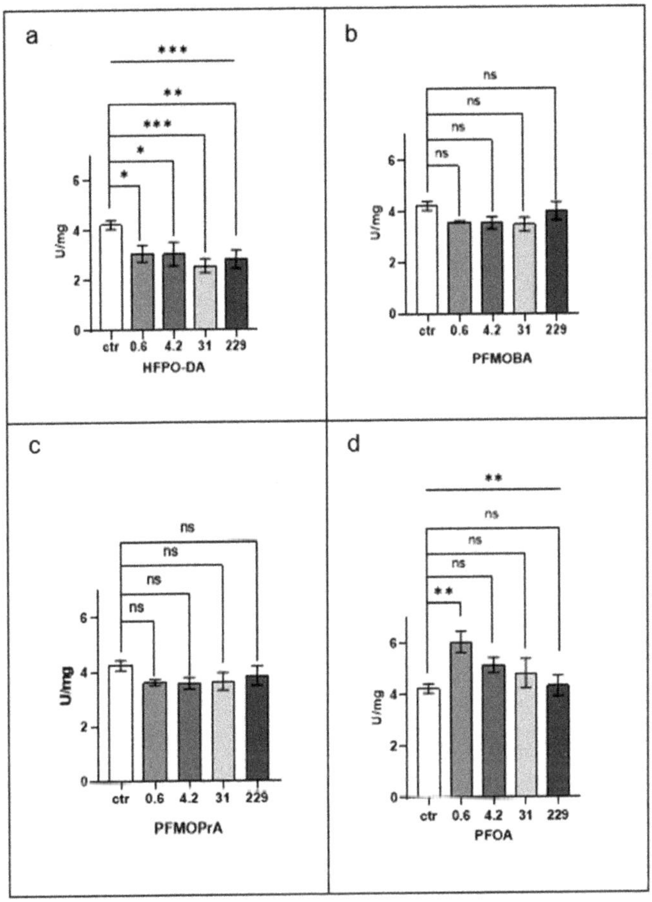

Figure 2. SOD (superoxide dismutase) activity. Statistical significance was determined by ANOVA followed by a post hoc Holm–Sidak's multiple comparison test. * $p < 0.05$; ** $p < 0.01$; and *** $p < 0.001$; ns, not statistically significant. The line above the histogram bars indicates statistical significance between PFAS-treated groups and the control. Each panel represents results for a different PFAS compound: (**a**) HFPO-DA, (**b**) PFMOBA, (**c**) PFMOPrA, and (**d**) PFOA. See caption to Figure 1 for more details.

3.2. Phenol Oxidase Activity

Figure 3 presents the changes in phenol oxidase (PO) activity in the haemolymph (serum) of *Eisenia fetida* after exposure to PFAS.

HFPO-DA exposure resulted in a significant reduction in PO activity compared to control levels at all concentrations tested. The most pronounced decrease was observed at 31 μM ($p < 0.001$).

PFMOBA exposure led to a significant decrease in PO activity only at the highest concentration of 229 μM ($p < 0.01$), while lower concentrations did not produce significant effects.

PFMOPrA demonstrated the strongest inhibitory effect on PO activity among the compounds tested, with significant reductions observed at all concentrations ($p < 0.0001$). This indicates a potent effect on the earthworm's immune function, as PO is crucial in defence mechanisms.

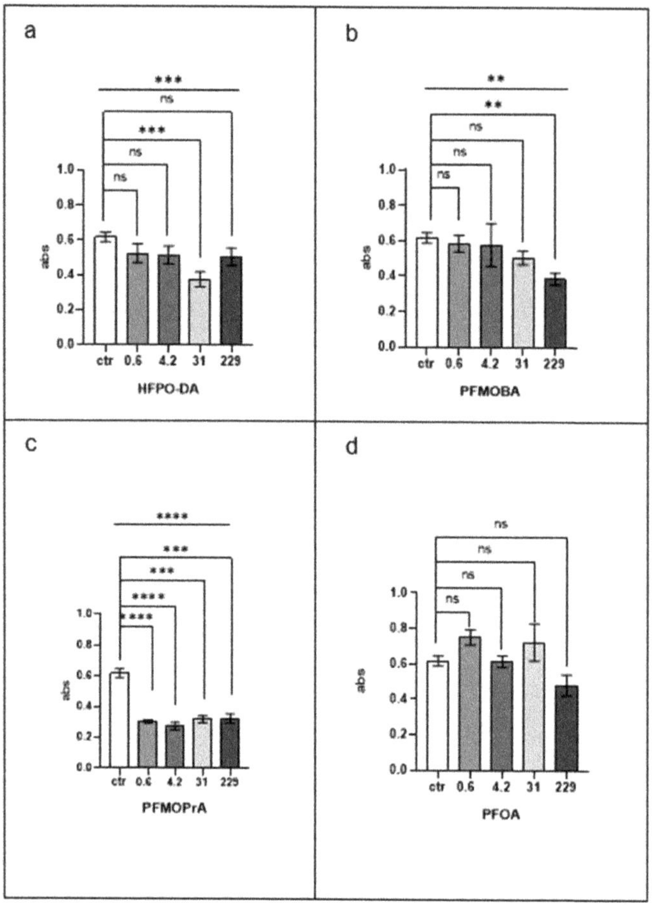

Figure 3. Phenol oxidase. A semi-quantitative evaluation was performed following the absorption (abs) at 590 nm due to L-DOPA oxidation. Statistical significance was determined by the Kruskal–Wallis test followed by a post hoc Dunn's multiple comparison test. ** $p < 0.01$; *** $p < 0.001$; and **** $p < 0.0001$; ns, not statistically significant. The line above the histogram bars indicates statistical significance between PFAS-treated groups and the control. Each panel represents results for a different PFAS compound: (**a**) HFPO-DA, (**b**) PFMOBA, (**c**) PFMOPrA, and (**d**) PFOA. See caption to Figure 1 for more details.

PFOA exposure did not significantly affect PO activity at any concentration, suggesting it does not impact this immune parameter.

PFMOPrA had the most substantial inhibitory effect on PO activity, followed by HFPO-DA. PFMOBA's effect was significant only at the highest concentration, and PFOA did not affect PO activity.

These results suggest that the emerging PFPOA compounds, particularly PFMOPrA and HFPO-DA, may impair immune function in earthworms by inhibiting PO activity, whereas PFOA lacks this effect.

3.3. *Acetylcholinesterase Activity*

Figure 4 depicts the effects of PFAS exposure on acetylcholinesterase (AChE) activity in *Eisenia fetida* tissues.

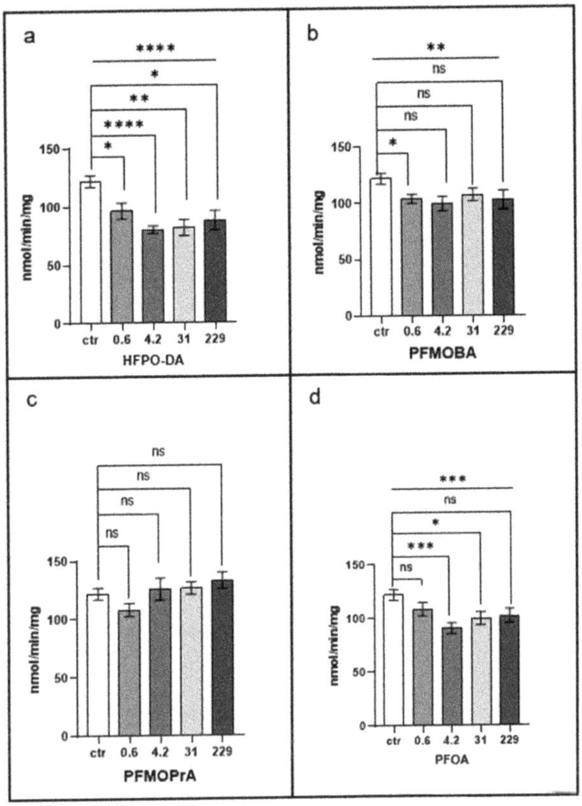

Figure 4. Acetylcholinesterase activity. Statistical significance was determined by the Brown–Forsythe ANOVA test, followed by a post hoc Dunnett's T3 test. * $p < 0.05$; ** $p < 0.01$; *** $p < 0.001$; and **** $p < 0.0001$; ns, not statistically significant. The line above the histogram bars indicates statistical significance between PFAS-treated groups and the control. Each panel represents results for a different PFAS compound: (**a**) HFPO-DA, (**b**) PFMOBA, (**c**) PFMOPrA, and (**d**) PFOA. See caption to Figure 1 for more details.

HFPO-DA exposure resulted in a significant decrease in AChE activity across multiple concentrations. Significant reductions were observed at 0.6 μM ($p < 0.01$), 4.2 μM ($p < 0.001$), 31 μM ($p < 0.0001$), and 229 μM ($p < 0.05$), indicating a dose-dependent inhibitory effect.

PFOA exposure led to a moderate decrease in AChE activity that was statistically significant at 4.2 μM ($p < 0.05$), suggesting a potential inhibitory effect, though less pronounced than that of HFPO-DA.

PFMOBA exposure exhibited threshold-like effects, with a significant reduction observed only at 229 μM ($p < 0.05$).

PFMOPrA did not significantly affect AChE activity at any concentration tested, indicating a lack of effect on this enzyme.

Comparing the compounds, HFPO-DA had the most robust inhibitory effect on AChE activity, showing significant decreases across all concentrations. PFOA and PFMOBA exhibited moderate inhibitory effects at specific concentrations, while PFMOPrA did not affect AChE activity.

These findings suggest that HFPO-DA may interfere with cholinergic neurotransmission in earthworms by inhibiting AChE activity, which could impact neuromuscular function.

3.4. Escape Test

The behavioural effects of PFAS exposure were evaluated using an escape latency test, with results presented in Figure 5.

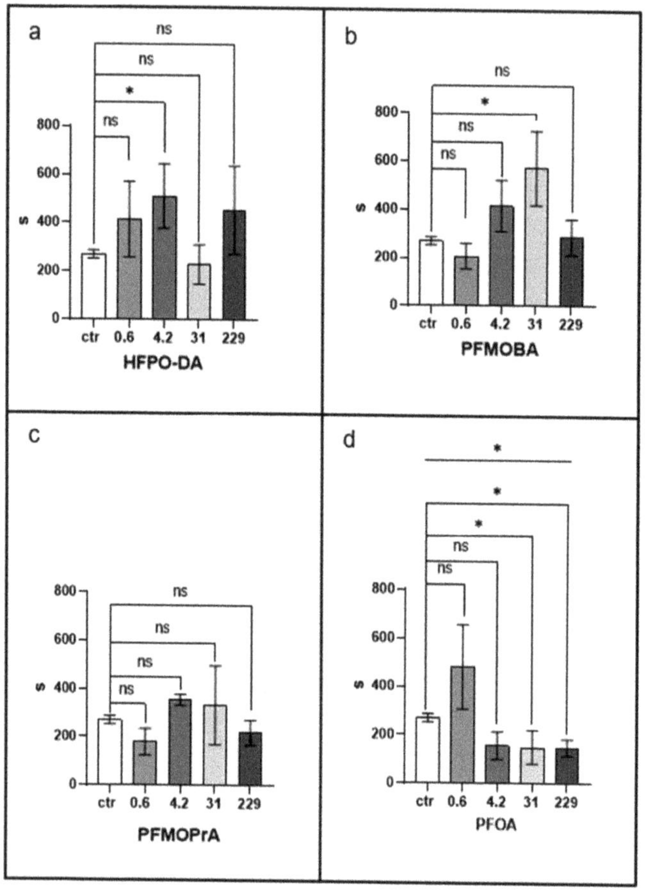

Figure 5. Escape test. Statistical significance was determined by the Kruskal–Wallis test, followed by a post hoc uncorrected Dunn's test. * $p < 0.05$; ns, not statistically significant. The line above the histogram bars indicates statistical significance between PFAS-treated groups and the control. Each panel represents results for a different PFAS compound: (**a**) HFPO-DA, (**b**) PFMOBA, (**c**) PFMOPrA, and (**d**) PFOA. See caption to Figure 1 for more details.

PFOA exposure resulted in a significant reduction in median escape latency compared to the control group ($p < 0.05$), indicating a quicker escape response. This effect was more consistent across the higher concentrations tested.

HFPO-DA and PFMOBA exposures led to increases in escape latency times, with some specimens exhibiting prolonged responses. For HFPO-DA, several earthworms failed to escape within the full duration of the assay (20 min), resulting in a positively skewed distribution and higher mean latency times.

PFMOPrA exposure did not produce significant changes in median escape latency compared to the control, though the data distribution showed instances of both longer and shorter latency times.

The distribution of latency times for PFAS-exposed groups differed significantly from that of the control group (Kolmogorov–Smirnov test, $p < 0.05$), indicating that PFAS exposure affects the behavioural responses of earthworms (Table 1).

Table 1. Escape latency time statistics. Shown are descriptive statistics obtained for the different tested conditions. All time data are expressed in seconds.

	CTRL	HFPO-DA	PFMOBA	PFMOPrA	PFOA
No. of cases	28	28	28	28	28
Minimum	22	15	8	10	2
Maximum	518	1220	1220	1220	1220
Range	496	1205	1212	1210	1218
Sum	7551	11,265	10,349	7591	6560
Median	283	21	296	257	169
Arithmetic Mean	269.68	402.35	369.61	271.11	234.29
SE	25.17	70.51	55.8	45.36	54.35
95,% Lower Confidence Limit	218.03	257.65	255.11	178.04	122.78
95,% Upper Confidence Limit	321.33	547	484.11	364.17	345.8
Standard Deviation	133.2	373.1	295.28	240.01	287.58
Shapiro–Wilk Statistics	0.96	0.84	0.91	0.79	0.74
Shapiro–Wilk p-value	0.26	0.001	0.016	0.0001	0.0001

Figure 6 shows histograms of latency times, highlighting deviations from the symmetrical distribution observed in the control group. PFAS-treated groups exhibited non-normal distributions with positive skewness, suggesting the presence of more extreme values.

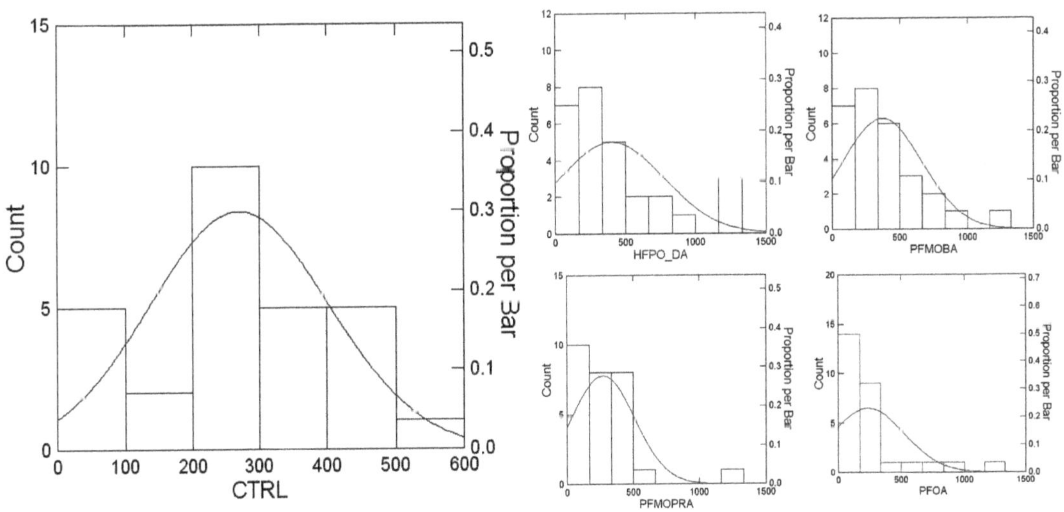

Figure 6. Histogram of latency times obtained from the escape test. The binned counts show the deviations of PFAS specimens from the symmetrical distribution of the control group. A biphasic response is observed, with a general increase in latency times for PFAS treatment. However, PFOA showed quicker median performance as judged by the U-statistics results (see body text).

PFOA exposure led to a faster escape response, while HFPO-DA and PFMOBA exposures resulted in delayed responses for some specimens. PFMOPrA had variable effects without significant changes in median latency.

These results suggest that PFAS compounds can affect the behavioural responses of earthworms, potentially impacting their ability to respond to adverse environmental conditions.

A Principal Component Analysis (PCA) was conducted using all data gathered in this study (Supplementary Figure S1). The PCA revealed separation among the treatments, suggesting distinct or partially overlapping modes of action and biochemical interactions for each PFAS congener. HFPO-DA and PFOA appeared clearly separated in the first and fourth quadrants of the loading plot, respectively. PFMOBA was intermediate, while PFMOPrA was separated in the third quadrant. The differences observed in the PCA were driven primarily by catalase and SOD activities in the first principal component (PC1) and acetylcholinesterase activity in the second principal component (PC2).

To complement the PCA analysis, an expert system classification was applied to compare the relative potency of the PFAS congeners based on five biochemical and behavioural endpoints: catalase, SOD, phenol oxidase, acetylcholinesterase activities, and escape latency. HFPO-DA ranked as the most potent compound, with the highest cumulative score (25), followed by PFMOPrA (20), PFMOBA (13), and PFOA (10). This classification, excluding PCA to avoid redundancy, highlighted HFPO-DA's consistent and broad toxicity profile, underscoring its significant ecotoxicological risk. Further details are provided in Supplementary Figure S2 and in Tables S1–S3 (see Supplementary Information).

4. Discussion

The accumulation of PFAS in the environment and their subsequent transfer through trophic chains have led to increased studies aimed at determining their impact on ecosystems. Ecotoxicological studies on well-characterised environmental models, such as earthworms, are valuable tools for assessing the potential toxicity of persistent contaminants like PFAS, focusing on their bioavailable fraction. Numerous studies report the accumulation of PFAS in various invertebrate organisms [6,8,33–36] and vertebrates [37–40]. The effects of PFAS on these organisms are diverse, varying according to their composition (with thousands of congeners), concentration, exposure time, and the species affected. Therefore, exposure to PFAS can lead to a range of consequences, from subtle biochemical changes to severe physiological disruptions.

One target of PFAS is cellular metabolism, and it has been observed that PFAS toxicity manifests at the molecular level by altering enzyme activities involved in metabolic pathways. This can affect growth, reproduction, and, in the most severe cases, the survival of the organism. In human hepatoma cells (HepG2), exposure to PFOA and PFOS resulted in an increase in the activity of superoxide dismutase (SOD) and catalase (CAT), as reported by Bonato et al. (2020) [41] in a review of findings on PFAS affecting oxidative stress biomarkers in vitro and in vivo. These studies showed that PFOA and PFOS can overwhelm the balance of the antioxidant system, increase ROS generation, affect mitochondria, and initiate apoptosis. Mitochondrial dysfunction has also been reported in zebrafish [42].

Our study provides critical insights into the biochemical and behavioural impacts of per- and polyfluoroalkyl substances (PFAS) on terrestrial oligochaetes, particularly *Eisenia fetida*. Given the limited data on PFAS effects in terrestrial organisms, our aim was to evaluate a broad spectrum of biochemical biomarkers, including parameters related to oxidative stress, immunotoxicity, and neurotoxicity, with a particular focus on behaviour. In our study, we found no mortality up to a concentration of 4.6 mM suggesting that acute toxicity is not the primary concern with PFAS, including next-generation congeners. This finding is consistent with previous findings with either legacy PFAS, i.e., no mortality up to 100 mg kg^{-1} [43], or alternative congeners such as C6O4, a perfluoroalkyl ether compound [16]. In the latter case, significant acute toxicity is reported only at high concentrations (NOEC, 1390 mg/kg dry weight (d.w.)) while highlighting more pronounced reproductive effects at lower levels (EC50: 10.4 mg/kg d.w.; EC10: 0.8 mg/kg d.w.). These results emphasise the potential for chronic toxicity due to sublethal impacts observed at sub-micromolar concentrations.

The first hypothesis of PFAS toxicity we tested was the superoxide theory, which can lead to mitochondrial dysfunction and neurotoxicity as shown in nematodes [44]. In our study, we observed a significant response of superoxide dismutase (SOD) activity in organisms exposed to PFOA, characterised by an initial increase at lower exposure levels followed by a return to basal levels. This pattern mirrors findings reported by Wang et al. (2021) [45] in terrestrial oligochaetes, where SOD activation was noted at low pollutant levels but declined at higher PFOA concentrations. Interestingly, our study also revealed a similar trend in earthworms exposed to HFPO-DA, with all exposure conditions (ranging from 0.6 to 229 μM) resulting in SOD inhibition. Additionally, while PFMOBA and PFMOPrA induced slight and non-significant alterations in superoxide dismutase activity in *Eisenia fetida*, these changes indicated a trend towards reduced enzyme function. The exact mechanisms behind these effects are difficult to pinpoint, but some authors have suggested the possibility of enzyme inhibition due to the accumulation of oxidising agents in the organism's body [45]. Moreover, it has been shown that certain PFAS congeners, such as PFOA and PFOS, can bind to the SOD enzyme, causing a change in its function [46,47]. The antioxidant defence system plays a crucial role in the toxicological interactions of PFOS and PFOA in *Daphnia magna* [46]. Exposure to PFAS appears to promote the accumulation of superoxide anions, which may induce the activation of the SOD synthesis pathway through mechanisms such as gene expression changes. The observed reduction in activity may also reflect a compensatory response following an initial enzyme peak or a metabolic effect at the level of gene or protein expression [48]. These factors could not be fully tested due to the limitations of our experimental design, which prioritised the number and concentrations of PFAS congeners in alignment with the 72 h period specified in OECD Test Guideline No. 207. However, the observed loss of SOD activity may indicate long-term toxicity of PFAS and therefore warrants further investigation in future studies.

The catalytic action of SOD is generally followed by catalase activity, which completes the detoxification process eliminating hydrogen peroxide from the cytoplasm. The various tests carried out in this study showed different responses of catalase between the legacy PFAS (PFOA) and the PFPOA congeners. In fact, the interesting aspect is that what has been observed with SOD is reciprocal to the effects on catalase: PFOA had no significant effects, whereas HFPO-DA, PFMOBA, and PFMOPrA appeared to stimulate catalase activity in *Eisenia fetida* in an increasing manner with C4 > C5 > C6. This means that these substances are likely to cause an increase in ROS concentration and a proactive detoxifying response. PFMOPrA induced a clear concentration-dependent increase in the detoxifying activity of CAT. In PFMOBA, this trend is also present with a decrease at the highest concentration, which is, however, an effect commonly described in the literature as the peak effect [49]. In fact, at the highest PFAS concentration (229 μM), excessive production of H_2O_2 might impair the antioxidant capacity of terrestrial oligochaetes, and, consequently, catalase was inhibited. Similar results have been reported by Zhao et al. (2017) [50], even though our study is based on a short-term filter paper exposure and theirs on a long-term approach in soil.

We then investigated the potential effects of new-generation PFAS compounds on the activity of phenol oxidase (PO), a key enzyme in the innate immune system. PO is crucial in the melanin cascade, promoting pathogen encapsulation and destruction. It is worth noting that there is a connection between superoxide and phenol oxidase activity. Both legacy and next-generation PFAS can alter the activity of antioxidant enzymes such as superoxide dismutase, catalase and peroxidase [41,51], which play an essential role in limiting the excessive formation of ROS by acting as a first line of defence against ROS-mediated tissue damage. Earthworm coelomocytes can synthesise melanin through the action of phenol oxidase, which is useful for encapsulating and dissociating pathogens from the organism. It should also be noted that highly reactive, toxic quinone intermediates are formed during this melanisation process [52]. Studies have shown that phenol oxidase has a fungicidal, bactericidal, and antiviral effect and is, therefore, an important component of the innate immune system of invertebrates [53]. In our study, a decrease in phenol oxidase activity was observed in PFAS-treated samples after exposure of earthworms to PFAS. The results of our investigation are consistent with previous studies that have demonstrated the detrimental effects of PFAS exposure on the activity of important enzymes involved in antioxidant defence mechanisms and innate immunity. The observed reduction in phenol oxidase activity indicates a potential disruption of the earthworms' ability to mount an effective immune response against pathogens. In mammals, PFAS exposure has been shown to alter CD4+ and CD8+ T cell populations, reducing their functionality and increasing pro-inflammatory cytokines like IL-1β, which compromises cell-mediated immunity [54]. Moreover, over the past decade, numerous epidemiological studies have highlighted the ability of PFASs to affect the immune system through immunosuppressive action, finding a decrease in antibody production with increasing PFAS levels in serum [55,56]. Our findings of disrupted PO activity in earthworms suggest a potential commonality in PFAS-induced immunotoxicity across different immune systems. This highlights the need for further studies to explore these parallels and inform comprehensive risk assessments of PFAS exposure.

AChE plays a very important role in the function of the neuromuscular system as it prevents continuous muscle contractions. Inhibition of AChE causes dysfunction in organisms, such as behavioural changes, paralysis, and death. AChE activity has been used to analyse soil contamination effects mediated by a broad spectrum of molecules, such as organophosphate pesticides, carbamates, persistent organic pollutants (POPs), and polycyclic aromatic hydrocarbons (PAHs) [22,57,58]. Mulkiewicz et al. (2007) [59] demonstrated AChE inhibitory effects of PFAS, i.e., perfluorocarboxylic acid congeners, including PFOA with IC50 in the mM range using an *in chemico* assay with the eel AChE enzyme. A study by Li (2007) [60] in *Dugesia planaria* showed in vivo AChE inhibition (about 30%) at PFOA concentrations similar to the highest dose levels adopted in our study. Other studies in planaria and Daphnia reported enzyme modulation due to PFOS [61]. A study by Nusair et al. (2017) [62] on *Eisenia andrei* showed that polychlorinated flame retardants may have the ability to block the action of AChE in earthworm specimens by 30%, but no effects of PFAS on oligochaete have been reported so far. Nevertheless, Yang et al. (2019) [46] proposed a mechanism of inhibition due to physical interactions of PFAS and the enzyme molecule, as shown by docking studies using genuine *D. magna* AChE-inferred protein sequence. In our study, the effect found in tests conducted on HFPO-DA is very similar to that of PFOA. In contrast, PFMOPrA and PFMOBA show results that are of little or no statistical significance. The signalling mechanisms for the inhibition of AChE by PFAS are still unclear and under investigation. For other compounds, however, they appear to be based on regulation at the gene expression level. The fact that the differences in AChE activity were not extreme suggests that these chemicals do not have a specific effect on the catalytic site of the enzyme. However, the effects observed on HFPO-DA and PFOA suggest that these chemicals do influence the activity of AChE.

Previous work has shown that enzymatic inhibition of AChE and increased acetylcholine synthesis also affect behaviour [62]. In a recent study on *Eisenia fetida* exposed to polycyclic aromatic hydrocarbons such as fluorene, He et al. (2023) [63] reported how altered AChE signalling affects avoidance behaviour, i.e., the organisms tested are less efficient at perceiving the presence of abiotic stress and also show altered growth and metabolism. Moreover, studies on male mice exposed to PFOS during lactation showed alterations in locomotor behaviour; in fact, treated mice were less active at the beginning of the experiment and had high activity at the end [64]. These animals also showed a reduced transcription of the gene for AChE. Experiments concerning PFAS-induced neurotoxicity mostly involve vertebrate animals, while studies on invertebrates are very few, and one of these is the one carried out on *C. elegans* [65] subjected to PFAS treatment, in which the animals showed clear locomotor deficits. In our study, results show that PFAS exposure, including the relatively newer PFPOA congeners, exerts a possible neuromotor deregulation in *Eisenia fetida*, and this effect has often been associated with AChE inhibition, a common marker of neurotoxicity found to be affected by many organic compounds including dioxins [66]. Behavioural responses, assessed through the escape test [34], revealed statistically significant effects of PFOA on escape latency, with notable deviations from the control group. This observation aligns with the findings in *Lumbriculus variegatus*, where PFOS exposure significantly affected escape behaviour [67]. The data distribution for the propylene oxide acids showed significant deviations from the control group, with increased latency times suggesting neuromotor deregulation. These behavioural alterations might be linked to the neurotoxic effects observed through decreased AChE activity. Mechanistic studies on the (neuro)toxicity of new-generation PFASs such as HFPO-DA are still scarce in the literature. To our knowledge, the only data found in the literature are from an experiment conducted on zebrafish larvae exposed to HFPO-DA treatment from day 0 to day 5 post-fertilisation [68], where HFPO-DA did not elicit neurotoxicity during the developmental stages of this organism. Despite this negative evidence, further measuring of neurotransmitter levels in earthworm tissues might provide further clues about the observed effects on locomotion.

This study sheds light on the sublethal impacts of short-chain PFPOA congeners, which are currently under-investigated, on terrestrial invertebrates. This research reveals that the sublethal effects of these compounds are evident and correlate with their specific biochemical and behavioural profiles, as indicated by the comparative potency analysis. The results demonstrate that HFPO-DA (C6) exerts the most pronounced effects, followed by PFMOPrA (C4) and PFMOBA (C5), with PFOA showing the least impact. This highlights the importance of considering both molecular structure and functional endpoints when assessing the ecotoxicological risks of emerging PFAS compounds.

The hypothesis that HFPO-DA (C6) is the most active molecule among the tested PFAS is supported by its impact on a wide range of measured variables. HFPO-DA significantly increased catalase activity at both high (229 µM) and low (0.6 µM) concentrations, indicating a significant influence on oxidative stress response. It consistently decreased SOD activity across all concentrations, suggesting a disruption of the antioxidant defence system. Furthermore, HFPO-DA significantly reduced phenol oxidase activity at 31 µM, potentially impairing the immune response. Remarkably, it also showed the most substantial inhibitory effect on AChE activity among the tested PFAS, implying a pronounced delay time in the escape test. These findings, along with the PCA plot indicating a distinct separation of HFPO-DA from other congeners and the expert system classification (Supplementary Information), collectively suggest that HFPO-DA is the most potent disruptor of the measured biological parameters in *E. fetida*.

One possible explanation for HFPO-DA's heightened potency is its unique hexafluoropropylene oxide functional group, which may enhance its ability to integrate into and disrupt biological membranes. This structural feature could facilitate interactions with key enzymes or proteins, leading to broader biochemical impacts. Additionally, HFPO-DA's physicochemical properties, such as its optimal balance of hydrophobicity and hydrophilicity, may allow it to penetrate cell membranes more effectively, enabling greater access to intracellular targets. In contrast, PFMOPrA's smaller size and simpler methoxy-functionalised structure may enhance its mobility and interaction with specific targets compared to PFMOBA, which is slightly larger and potentially less bioavailable. These structural differences provide a plausible basis for the observed ranking in toxicity and highlight the nuanced role of molecular structure in PFAS bioactivity.

This study bears limitations due to its short-term nature and would benefit from long-term investigations. However, it showed remarkable effects on vital aspects of earthworm metabolism, including antioxidant, immune, neurological, and behavioural functions, starting from concentrations that are commonly found in hotspots of PFAS contamination (0.6 µM PFOA equals nearly 250 ppb), such as firefighting sites and manufacturing districts. Our study reveals significant biochemical, immunotoxic, neurological, and behavioural effects of next-generation PFAS on earthworms at sub-micromolar levels. These findings highlight a potential threat to soil health and biodiversity. Earthworms play a crucial role in soil ecosystems by aiding nutrient cycling, organic matter decomposition, and maintaining soil structure. The observed adverse effects indicate that PFAS contamination can disrupt these essential ecosystem services, leading to broader ecological consequences. Disruption of earthworm physiological functions can impair soil aeration, nutrient cycling, and organic matter decomposition, potentially affecting soil fertility and plant growth. Moreover, earthworms serve as prey for a variety of animals, so adverse effects on earthworm populations could have cascading impacts on food webs.

By highlighting the sublethal effects of next-generation PFAS, our study underscores the importance of including terrestrial organisms in environmental risk assessments. Our results suggest that these compounds warrant closer scrutiny to understand their potential long-term impacts on soil ecosystems. Further research is needed to investigate the chronic toxicity of PFPOAs, including reproductive and developmental endpoints, and to explore the molecular mechanisms underlying the observed effects.

5. Conclusions

This study highlights the complex and varying impacts of different PFPOAs on terrestrial invertebrates, emphasising that while acute toxicity appears low, chronic exposure to even sublethal levels of these substances could result in significant physiological and behavioural disruptions. The observed effects on antioxidant enzymes, immune function, neurotoxicity markers, and behaviour underline the multifaceted impact of PFAS. Furthermore, the comparative potency analysis conducted in this study revealed that HFPO-DA exhibits the most pronounced effects, followed by PFMOPrA, PFMOBA, and PFOA, demonstrating the need for comprehensive ecotoxicological assessments that integrate biochemical and behavioural endpoints alongside systematic potency classification frameworks.

Supplementary Materials: The following supporting information can be downloaded at https://www.mdpi.com/article/10.3390/jox15010002/s1, Figure S1: Principal component analysis on standardised data; Figure S2: PFPOA potency vs. PFOA; Table S1: PFAS congener potency summary; Table S2: Scoring system; Table S3: Scoring system.

Author Contributions: D.R.: conceptualization, methodology, investigation, formal Analysis, data Curation, statistical analysis, software, writing—original draft, review, editing, visualization. D.G.: methodology, formal analysis, statistical analysis, visualization. C.L.: review, methodology visualization. A.B.: editing, visualization. A.C.: review, methodology, investigation. F.D.: conceptualization, methodology, investigation, validation, formal analysis, data curation, statistical analysis, writing—original draft, review, editing, visualization, project administration, funding acquisition. All authors have read and agreed to the published version of the manuscript.

Funding: This project has received funding from the European Union's Horizon 2020 research and innovation programme under grant agreement No 101037509 (SCENARIOS project).

Institutional Review Board Statement: Not applicable.

Informed Consent Statement: Not applicable.

Data Availability Statement: The original contributions presented in this study are included in the article. Further inquiries can be directed to the corresponding author.

Acknowledgments: We sincerely thank the anonymous reviewers for their valuable comments and constructive feedback, which have significantly improved the quality and clarity of this manuscript.

Conflicts of Interest: The authors declare no conflicts of interest.

References

1. Pelch, K.E.; Reade, A.; Wolffe, T.A.M.; Kwiatkowski, C.F. PFAS health effects database: Protocol for a systematic evidence map. *Environ. Int.* **2019**, *130*, 104851. [CrossRef] [PubMed]
2. Houde, M.; Martin, J.W.; Letcher, R.J.; Solomon, K.R.; Muir, D.C. Biological monitoring of polyfluoroalkyl substances: A review. *Environ. Sci. Technol.* **2006**, *40*, 3463–3473. [CrossRef]
3. Brase, R.A.; Mullin, E.J.; Spink, D.C. Legacy and emerging per- and polyfluoroalkyl substances: Analytical techniques, environmental fate, and health effects. *Int. J. Mol. Sci.* **2021**, *22*, 995. [CrossRef]
4. Van Asselt, E.D.; Rietra, R.P.; Römkens, P.F.; van der Fels-Klerx, H.J. Perfluorooctane sulphonate (PFOS) throughout the food production chain. *Food Chem.* **2011**, *128*, 1–6. [CrossRef]
5. Rich, C.D.; Blaine, A.C.; Hundal, L.; Higgins, C.P. Bioaccumulation of perfluoroalkyl acids by earthworms (*Eisenia foetida*) exposed to contaminated soils. *Environ. Sci. Technol.* **2015**, *49*, 881–888. [CrossRef] [PubMed]
6. Burkhard, L.P.; Votava, L.K. Biota-Sediment Accumulation Factors for Per- and Polyfluoroalkyl Substances. *Environ. Toxicol. Chem.* **2023**, *42*, 277–295. [CrossRef] [PubMed]
7. Brusseau, M.L.; Anderson, R.H.; Guo, B. PFAS concentrations in soils: Background levels versus contaminated sites. *Sci. Total Environ.* **2020**, *740*, 140017. [CrossRef]
8. Zhao, S.; Zhu, L.; Liu, L.; Liu, Z.; Zhang, Y. Bioaccumulation of perfluoroalkyl carboxylates (PFCAs) and perfluoroalkane sulfonates (PFSAs) by earthworms (*Eisenia foetida*) in soil. *Environ. Pollut.* **2013**, *179*, 45–52. [CrossRef] [PubMed]
9. Zhao, L.; Zhu, L.; Zhao, S.; Ma, X. Sequestration and bioavailability of perfluoroalkyl acids (PFAAs) in soils: Implications for their underestimated risk. *Sci. Total. Environ.* **2016**, *572*, 169–176. [CrossRef]
10. Ghisi, R.; Vamerali, T.; Manzetti, S. Accumulation of perfluorinated alkyl substances (PFAS) in agricultural plants: A review. *Environ. Res.* **2019**, *169*, 326–341. [CrossRef] [PubMed]
11. Shahsavari, E.; Rouch, D.; Khudur, L.S.; Thomas, D.; Aburto-Medina, A.; Ball, A.S. Challenges and current status of the biological treatment of pfas-contaminated soils. *Front. Bioeng. Biotechnol.* **2021**, *8*, 602040. [CrossRef] [PubMed]
12. Li, J.; Sun, J.; Li, P. Exposure routes, bioaccumulation and toxic effects of per- and polyfluoroalkyl substances (PFASs) on plants: A critical review. *Environ. Int.* **2022**, *158*, 106891. [CrossRef]
13. Wang, S.; Liu, T.; Qian, X.; Wang, H.; Li, M.; Wang, X.; Wei, S.; Chen, H. Microbial plankton responses to perfluoroalkyl acids and their alternatives in the aquatic environment. *J. Hazard. Mat.* **2023**, *441*, 129980. [CrossRef]
14. Zhao, S.; Fang, S.; Zhu, L.; Liu, L.; Liu, Z.; Zhang, Y. Mutual impacts of wheat (*Triticum aestivum* L.) and earthworms (*Eisenia fetida*) on the bioavailability of perfluoroalkyl substances (PFASs) in soil. *Environ. Pollut.* **2014**, *184*, 495–501. [CrossRef]
15. Karnjanapiboonwong, A.; Deb, S.K.; Subbiah, S.; Wang, D.; Anderson, T.A. Perfluoroalkylsulfonic and carboxylic acids in earthworms (*Eisenia fetida*): Accumulation and effects results from spiked soils at PFAS concentrations bracketing environmental relevance. *Chemosphere* **2018**, *199*, 168–173. [CrossRef] [PubMed]
16. Bizzotto, E.C.; Libralato, G.; Breda, S.; Siciliano, A.; Scanferla, P.; Vighi, M.; Marcomini, A. Toxicity and bioaccumulation of the fluorosurfactant cC6O4 in the earthworm *Eisenia foetida* (Savigny, 1826). *Sci. Total Environ.* **2024**, *919*, 170677. [CrossRef]
17. Calisi, A.; Baranzini, N.; Marcolli, G.; Bon, C.; Rotondo, D.; Gualandris, D.; Pulze, L.; Grimaldi, A.; Dondero, F. Evaluation of per- and polyfluoroalkyl substances (PFAS) toxic effects on the acute inflammatory response in the medicinal leech Hirudo verbana. *Chemosphere* **2024**, *366*, 143519. [CrossRef] [PubMed]
18. Zahm, S.H.; Bonde, J.P.; Chiu, W.A.; Hoppin, J.A.; Kanno, J.; Abdallah, M.; Blystone, C.R.; Calkins, M.; Dong, G.; Dorman, D.C.; et al. Carcinogenicity of perfluorooctanoic acid and perfluorooctanesulfonic acid. *Lancet Oncol.* **2024**, *25*, 16–17. [CrossRef]
19. Xu, D.; Li, C.; Wen, Y.; Liu, W. Antioxidant defense system responses and DNA damage of earthworms exposed to Perfluorooctane sulfonate (PFOS). *Environ. Pollut.* **2013**, *174*, 121–127. [CrossRef]
20. Delor, L.; Louzon, M.; Pelosi, C.; Michel, E.; Maillet, G.; Carronnier, H. Ecotoxicity of single and mixture of perfluoroalkyl substances (PFOS and PFOA) in soils to the earthworm *Aporrectodea caliginosa*. *Environ. Pollut.* **2023**, *335*, 122221. [CrossRef] [PubMed]

21. Munoz, G.; Desrosiers, M.; Vetter, L.; Duy, S.V.; Jarjour, J.; Liu, J.; Sauvé, S. Bioaccumulation of zwitterionic polyfluoroalkyl substances in earthworms exposed to aqueous Film-Forming foam impacted soils. *Environ. Sci. Technol.* **2020**, *54*, 1687–1697. [CrossRef] [PubMed]
22. Calisi, A.; Latino, M.E.; Corallo, A.; Grimaldi, A.; Ferronato, C.; Antisari, L.V.; Dondero, F. Biomarkers in Soil Organisms: Their Potential use in the Assessment of Soil Pollution and Remediation. In *Bioremediation of Agricultural Soils*; Sanchez-Hernandez, J.C., Ed.; CRC Press: Boca Raton, FL, USA, 2019; pp. 262–288.
23. Sanchez-Hernandez, J.C. *Bioremediation of Agricultural Soils*, 1st ed.; CRC Press: Boca Raton, FL, USA, 2019.
24. Xiao, R.; Ali, A.; Xu, Y.; Abdelrahman, H.; Li, R.; Lin, Y.; Bolan, N.; Shaheen, S.; Rinklebe, J.; Zhang, Z. Earthworms as candidates for remediation of potentially toxic elements contaminated soils and mitigating the environmental and human health risks: A review. *Environ. Int.* **2022**, *158*, 106924. [CrossRef]
25. Alves, P.R.L.; Bandeira, F.O.; Hennig, T.B. Ecological Role of Earthworms as Bioindicators of Soil Health. In *Earthworm Engineering and Applications*; Vig, A.P., Singh, J., Suthar, S.S., Eds.; Nova Science Publishers: New York, NY, USA, 2022; pp. 55–81.
26. Sun, M.; Arevalo, E.; Strynar, M.; Lindstrom, A.; Richardson, M.; Kearns, B.; Pickett, A.; Smith, C.; Knappe, D.R.U. Legacy and emerging perfluoroalkyl substances are important drinking water contaminants in the Cape Fear River watershed of North Carolina. *Environ. Sci. Technol. Lett.* **2016**, *3*, 415–419. [CrossRef]
27. OECD. *Test No. 207: Earthworm, Acute Toxicity Tests, OECD Guidelines for the Testing of Chemicals, Section 2*; OECD Publishing: Paris, France, 1984. [CrossRef]
28. Zhang, C.; McElroy, A.C.; Liberatore, H.K.; Alexander, N.L.M.; Knappe, D.R.U. Stability of Per- and Polyfluoroalkyl Substances in Solvents Relevant to Environmental and Toxicological Analysis. *Environ. Sci. Technol.* **2022**, *56*, 6103–6112. [CrossRef] [PubMed] [PubMed Central]
29. Riedl, S.a.B.; Völkl, M.; Holzinger, A.; Jasinski, J.; Jérôme, V.; Scheibel, T.; Feldhaar, H.; Freitag, R. In vitro cultivation of primary intestinal cells from Eisenia fetida as basis for ecotoxicological studies. *Ecotoxicology* **2021**, *31*, 221–233. [CrossRef] [PubMed]
30. Lorusso, C.; Calisi, A.; Sarà, G.; Dondero, F. In-gel assay to evaluate antioxidant enzyme response to silver nitrate and silver nanoparticles in marine bivalve tissues. Special Issue: Aquatic Toxicology and Aquaculture: Questions and Advances. *Appl. Sci.* **2022**, *12*, 2760. [CrossRef]
31. Prochazkova, P.; Roubalova, R.; Skanta, F.; Dvorak, J.; Pacheco, N.I.N.; Kolarik, M.; Bilej, M. Developmental and immune role of a novel multiple cysteine cluster tlr from *Eisenia andrei* earthworms. *Front. Immunol.* **2019**, *10*, 1277. [CrossRef] [PubMed]
32. Da Cruz Jung, I.E.; Assmann, C.E.; Mastella, M.H.; Barbisan, F.; Spilliari Ruaro, R.A.; Roggia, I.; Osmarin Turra, B.; Chitolina, B.; de Oliveira Alves, A.; Teixeira, C.F.; et al. Superoxide-anion triggers impairments of immune efficiency and stress response behaviors of *Eisenia fetida* earthworms. *Chemosphere* **2021**, *269*, 128712. [CrossRef] [PubMed]
33. Mommaerts, V.; Hagenaars, A.; Meyer, J.; De Coen, W.; Swevers, L.; Mosallanejad, H.; Smagghe, G. Impact of a perfluorinated organic compound PFOS on the terrestrial pollinator *Bombus terrestris* (Insecta, Hymenoptera). *Ecotoxicology* **2011**, *20*, 447–456. [CrossRef]
34. Marziali, L.; Rosignoli, F.; Valsecchi, S.; Polesello, S.; Stefani, F. Effects of perfluoralkyl substances on a multigenerational scale: A case study with *Chironomus riparius* (diptera, chironomidae). *Chemosphere* **2019**, *38*, 988–999. [CrossRef]
35. Sonter, C.A.; Rader, R.; Stevenson, G.; Stavert, J.R.; Wilson, S.C. Biological and behavioural responses of European honey bee (*Apis mellifera*) colonies to perfluorooctane sulfonate exposure. *Integr. Environ. Assess. Manag.* **2021**, *17*, 673–683. [CrossRef]
36. Ma, T.; Ye, C.; Wang, T.; Li, X.; Luo, Y. Toxicity of per- and polyfluoroalkyl substances to aquatic invertebrates, planktons, and microorganisms. *Int. J. Environ. Res.* **2022**, *19*, 16729. [CrossRef] [PubMed]
37. Riebe, R.A.; Falk, S.; Georgii, S.; Brunn, H.; Failing, K.; Stahl, T. Perfluoro-alkyl acid concentrations in livers of fox (*Vulpes vulpes*) and chamois (*Rupicapra rupicapra*) from Germany and Austria. *Arch. Environ. Contam. Toxicol.* **2016**, *71*, 7–15. [CrossRef]
38. Babut, M.; Labadie, P.; Simonnet-Laprade, C.; Munoz, G.; Roger, M.-C.; Ferrari BJ, D.; Budzinski, H.; Sivade, E. Per- and poly-fluoroalkyl compounds in freshwater fish from the Rhône River: Influence of fish size, diet, prey contamination and biotransformation. *Sci. Total Environ.* **2017**, *605–606*, 38–47. [CrossRef] [PubMed]
39. Cleary, R.S.; Karnjanapiboonwong, A.; Thompson, W.A.; Lasee, S.J.; Subbiah, S.; Kaublé, R.K.; Andraski, B.J.; Anderson, T.A. Emerging and Historical Con-taminants Detected in Desert Rodents Collected Near a Low-Level Radioactive Waste Site. *Environ. Toxicol. Chem.* **2020**, *40*, 727–734. [CrossRef] [PubMed]
40. Abercrombie, S.A.; de Perre, C.; Iacchetta, M.; Flynn, R.W.; Sepúlveda, M.S.; Lee, L.S.; Hoverman, J.T. Sublethal Effects of Dermal Exposure to Poly- and Perfluoroalkyl Substances on Post metamorphic Amphibians. *Environ. Toxicol. Chem.* **2020**, *40*, 717–726. [CrossRef]
41. Bonato, M.; Corrà, F.; Bellio, M.; Guidolin, L.; Tallandini, L.; Irato, P.; Santovito, G. PFAS Environmental Pollution and Antioxidant Responses: An Overview of the Impact on Human Field. *Int. J. Environ. Res. Public Health* **2020**, *17*, 8020. [CrossRef]
42. Ulhaq, Z.S.; Tse, W.K. Perfluorohexanesulfonic acid (PFHxS) induces oxidative stress causes developmental toxicities in zebrafish embryos. *J. Hazard. Mat.* **2023**, *457*, 131722. [CrossRef] [PubMed]

43. He, W.; Megharaj, M.; Naidu, R. Toxicity of perfluorooctanoic acid towards earthworm and enzymatic activities in soil. *Environ. Monit. Assess.* **2016**, *188*, 424. [CrossRef]
44. Sammi, S.R.; Foguth, R.M.; Nieves, C.S.; De Perre, C.; Wipf, P.; McMurray, C.T.; Lee, L.S.; Cannon, J.R. Perfluorooctane Sulfonate (PFOS) Produces Dopaminergic Neuropathology in Caenorhabditis elegans. *Toxicol. Sci.* **2019**, *172*, 417–434. [CrossRef] [PubMed]
45. Wang, Z.; Li, C.; Shao, Y.; Xue, W.; Wang, N.; Xu, X.; Zhang, Z. Antioxidant defense system responses, lysosomal membrane stability and DNA damage in earthworms (*Eisenia fetida*) exposed to perfluorooctanoic acid: An integrated biomarker approach to evaluating toxicity. *RSC Adv.* **2021**, *11*, 26481–26492. [CrossRef] [PubMed]
46. Yang, H.B.; Zhao, Y.Z.; Tang, Y.; Gong, H.Q.; Guo, F.; Sun, W.H.; Liu, S.S.; Tan, H.; Chen, F. Antioxidant defence system is responsible for the toxicological interactions of mixtures: A case study on PFOS and PFOA in *Daphnia magna*. *Sci. Total. Environ.* **2019**, *667*, 435–443. [CrossRef]
47. Rajak, P.; Ganguly, A. The ligand-docking approach explores the binding affinity of PFOS and PFOA for major endogenous antioxidants: A potential mechanism to fuel oxidative stress. *Sust. Chem. Environ.* **2023**, *4*, 100047. [CrossRef]
48. Mayilswami, S.; Krishnan, K.; Megharaj, M.; Naidu, R. Chronic PFOS exposure alters the expression of neuronal development-related human homologues in *Eisenia fetida*. *Ecotoxicol. Environ. Saf.* **2014**, *110*, 288–297. [CrossRef]
49. Keillor, J.W.; Brown, R.S. Attack of zwitterionic ammonium thiolates on a distorted anilide as a model for the acylation of papain by amides. A simple demonstration of a bell-shaped pH/rate profile. *J. Am. Chem. Soc.* **1992**, *114*, 7983–7989. [CrossRef]
50. Zhao, Y.; Li, G.; Qi, D.; Sun, L.; Wen, C.; Yin, S. Biomarker responses of earthworms (*Eisenia fetida*) to soils contaminated with perfluorooctanoic acid. *Environ. Sci. Pollut. Res.* **2017**, *24*, 22073–22081. [CrossRef]
51. Briels, N.; Ciesielski, T.M.; Herzke, D.; Jaspers, V.L.B. Developmental toxicity of perfluorooctanesulfonate (PFOS) and its chlorinated polyfluoroalkyl ether sulfonate alternative F-53B in the domestic chicken. *Environ. Sci. Technol.* **2018**, *52*, 12859–12867. [CrossRef]
52. Cerenius, L.; Söderhäll, K. The prophenoloxidase-activating system in invertebrates. *Immunol. Rev.* **2004**, *198*, 116–126. [CrossRef]
53. Nappi, A.J.; Ottaviani, E. Cytotoxicity and cytotoxic molecules in invertebrates. *BioEssays* **2000**, *22*, 469–480. [CrossRef]
54. Ehrlich, V.; Bil, W.; Vandebriel, R.; Granum, B.; Luijten, M.; Lindeman, B.; Grandjean, P.; Kaiser, A.M.; Hauzenberger, I.; Hartmann, C.; et al. Consideration of pathways for immunotoxicity of per-and polyfluoroalkyl substances (PFAS). *Environ. Health* **2023**, *22*, 19. [CrossRef] [PubMed]
55. Crawford, L.; Halperin, S.A.; Dzierlenga, M.W.; Skidmore, B.; Linakis, M.W.; Nakagawa, S.; Longnecker, M.P. Systematic review and meta-analysis of epidemiologic data on vaccine response in relation to exposure to five principal perfluoroalkyl substances. *Environ. Int.* **2023**, *172*, 107734. [CrossRef]
56. DeWitt, J.C.; Blossom, S.J.; Schaider, L.A. Exposure to per-fluoroalkyl and polyfluoroalkyl substances leads to immunotoxicity: Epidemiological and toxicological evidence. *J. Expo. Sci. Environ. Epidemiol.* **2019**, *229*, 148–156. [CrossRef]
57. Lionetto, M.G.; Caricato, R.; Calisi, A.; Giordano, M.E.; Schettino, T. Acetylcholinesterase as biomarkers in environmental and occupational medicine: New insights and future perspectives. *Biomed. Res. Int.* **2013**, *2013*, 321313. [CrossRef]
58. Dondero, F.; Calisi, A. Evaluation of Pollution Effects in Marine Organisms: "Old" and "New Generation" Biomarkers. In *Coastal Ecosystems: Experiences and Recommendations for Environmental Monitoring Programs*; Sebastià, M.T., Ed.; Nova Science Publishers: New York, NY, USA, 2015; pp. 143–192.
59. Mulkiewicz, E.; Jastorff, B.; Składanowski, A.; Kleszczyński, K.; Stepnowski, P. Evaluation of the acute toxicity of perfluorinated carboxylic acids using eukaryotic cell lines, bacteria and enzymatic assays. *Environ. Toxicol. Pharmacol.* **2006**, *23*, 279–285. [CrossRef]
60. Li, M.H. Effects of nonionic and ionic surfactants on survival, oxidative stress, and cholinesterase activity of planarian. *Chemosphere* **2008**, *70*, 1796–1803. [CrossRef]
61. Jeong, T.Y.; Yuk, M.S.; Jeon, J.; Kim, S.D. Multigenerational effect of perfluorooctane sulfonate (PFOS) on the individual fitness and population growth of *Daphnia magna*. *Sci. Total. Environ.* **2016**, *569*, 1553–1560. [CrossRef] [PubMed]
62. Nusair, S.D.; Abu Zarour, Y.S.; id Altarifi, A.A. Effects of dibenzo-p-dioxins/dibenzofurans on acetylcholinesterase activity and histopathology of the body wall of earthworm *Eisenia andrei*: A potential biomarker for ecotoxicity monitoring. *Water Air Soil Pollut.* **2017**, *228*, 266. [CrossRef]
63. He, F.; Liu, R.; Tian, G.; Qi, Y.; Wang, T. Ecotoxicological evaluation of oxidative stress-mediated neurotoxic effects, genetic toxicity, behavioral disorders, and the corresponding mechanisms induced by fluorene-contaminated soil targeted to earthworm (*Eisenia fetida*) brain. *Sci. Total. Environ.* **2023**, *871*, 162014. [CrossRef] [PubMed]
64. Brown-Leung, J.M.; Cannon, J.R. Neurotransmission Targets of Per- and Polyfluoroalkyl Substance Neurotoxicity: Mechanisms and Potential Implications for Adverse Neurological Outcomes. *Chem. Res. Toxicol.* **2022**, *35*, 1312–1333. [CrossRef]
65. Chen, N.; Li, J.; Li, D.; Yang, Y.; He, D. Chronic Exposure to Perfluorooctane Sulfonate Induces Behavior Defects and Neurotoxicity through Oxidative Dam-ages, In Vivo and In Vitro. *PLoS ONE* **2014**, *9*, e113453. [CrossRef]
66. Fu, H.; Xia, Y.; Chen, Y.; Xu, T.; Xu, L.; Guo, Z.; Xu, H.; Xie, H.Q.; Zhao, B. Acetylcholinesterase is a potential biomarker for a broad spectrum of organic environmental pollutants. *Environ Sci Technol.* **2018**, *52*, 8065–8074. [CrossRef] [PubMed]

67. Wang, N.; Jagani, R.; Nwobodo, N.; Ma, J. Toxicity of environmentally relevant concentration of PFAS chemicals in *Lumbriculus variegatus* (Oligochaeta, Lumbriculidae)—A multi-bioindicator study. *Ecotoxicol. Environ. Saf.* **2023**, *268*, 115722. [CrossRef] [PubMed]
68. Gaballah, S.; Swank, A.; Sobus, J.R.; Howey, X.M.; Schmid, J.; Catron, T.; Mc Cord, J.; Hines, E.; Strynar, M.; Tal, T. Evaluation of Developmental Toxicity, Developmental Neurotoxicity, and Tissue Dose in Zebrafish Exposed to GenX and Other PFAS. *Environ. Sci. Pollut. Res.* **2020**, *128*, 47005. [CrossRef] [PubMed]

Journal of *Xenobiotics*

MDPI

Article

Apis mellifera as a Model Species to Evaluate Toxicological Effects of Fungicides Used in Vineyard Agroecosystems

Tommaso Campani [1,*], Ginevra Manieri [1], Ilaria Caliani [1], Agata Di Noi [2] and Silvia Casini [1]

1 Department of Physical, Earth and Environmental Sciences, University of Siena, Via Mattioli 4, 53100 Siena, Italy
2 Santa Chiara Lab, Università di Siena, Via Val di Montone, 1, 53100 Siena, Italy
* Correspondence: campani@unisi.it

Abstract: Agroecosystems provide habitats, food, and water for many pollinators and insects, but they are also heavily exposed to threats from the widespread use of pesticides and fertilizers. Managed honeybees and wild bees encounter pesticides in vineyards by collecting morning dew from vine leaves and accessing sugars from grapes, particularly during dry periods. This study assessed the toxicological effects of the commercial fungicide formulations Fantic FNCWG® and Ramedit combi®, both individually and in combination, on honeybees. Using a multi-biomarker approach, we evaluated neurotoxicity, metabolic disturbances, phase II detoxification processes, and immune system function. Our findings revealed that commercial fungicide mixtures with multiple active ingredients affect bees differently than single active compounds. Biomarker responses highlighted how these complex mixtures disrupt various enzymatic pathways; including immune function; altering critical enzyme kinetics involved in detoxification and potentially impairing essential bee functions. This study emphasizes the need for more comprehensive research into the sublethal effects of commercial pesticides, particularly those used in vineyards, which are understudied compared to pesticides used in orchards.

Keywords: pollinators; pesticides; biomarkers; toxicological effects; agroecology

1. Introduction

The factors behind the loss of honeybees are many, and climate change, habitat destruction, the increase in diseases and parasites [1,2], and exposure to heavy metals and pesticides are the most important [3,4]. This last category of compounds is a threat of great relevance to bees' health, especially due to intensive agriculture, which led to a significant increase in the use of pesticides in recent years [5]. The study of the negative effects of plant protection products on honeybees has always focused on those substances that directly come into contact with them such as insecticides, widely used in orchards, fungicides and herbicides [6–8] and, more generally, the substances that act on flowers were bees feed. On the other hand, it is increasingly important to study the effects of substances widely used in crops where pollinators are not typically expected to be present, such as in vineyards [9].

Several European regions are famous for wine production, with many hectares of land covered by vineyards as a monoculture. During droughts, when blooms are scarce, vineyards become vital feeding grounds for bees, with honeybees gathering morning dew from vine leaves and drawing sugar and water from the grapes.

These behaviors are accentuated by continuous climate change and can cause the exposure of bees to a group of plant protection products little investigated. The main treatments applied in a grapevine are fungicides, among which copper and sulfur are most frequent [10], but synthetic compounds also find frequent applications [11]. Copper-based fungicides, such as Bordeaux mixture, are widely used to control the downy mildew disease of grapevine cultivation [12,13]. Copper has been used in viticulture for more than 150 years, at rates of up to 80 kg ha^{-1} per year, which has led to the accumulation of copper in the topsoil of many vineyards [14]. The use of copper fungicides in organic agriculture is currently restricted in Europe [15]. To date, the amount of copper intensively used in organic farming is limited to 4 kg ha^{-1} per year in most European countries, including Italy, France, and Spain [16]. Used in combination with copper, the cymoxanil is a broad-spectrum, systemic fungicide used to protect various fruits, vegetables, and field crops from a wide range of fungal diseases and is particularly effective in controlling downy mildew [17,18]. It belongs to a class of aliphatic nitrogen compounds and functions as a foliar fungicide with both protective and curative effects. The literature about the toxicological effects of cymoxanil on bees or other insects is absent, and the literature about the effects on vertebrates is scarce. According to a study by Ahmed et al. [19], cymoxanil significantly increased levels of alkaline S-transferase (AST), alanine transaminase (ALT), and alkaline phosphatase (ALP) in the serum and liver of mice, suggesting tissue necrosis and potential leakage of these enzymes into the bloodstream.

The work of Chang et al. [20] suggested that famoxadone-cymoxanil can induce developmental toxicity, immunotoxicity and neurobehavioral toxicity in zebrafish larvae.

Another fungicide widely used since the fifties in the vineyard is folpet. Folpet is an organochlorine phthalimide with a molecular weight of 296.6 g/mol and is used as a protective, broad-spectrum fungicide against leaf spot diseases in grapevines. It acts by reacting with thiols, altering proteins and enzymes in fungi [21]. By nonspecific reaction with thiols, folpet reacts with cysteine amino acids in proteins and glutathione, thus affecting the function of many proteins and enzymes, altering honeybee worker gene expression and disturbing energy production with a reduction in foraging activity and hormonal dysregulation, such as the transition of nurse bees to foragers [22]. Moreover, folpet was found to induce cytotoxicity and tumorigenic activity in the gastrointestinal tract in mice [23] and mortality in larval stages of *Rana temporaria* and *Bufotes viridis* [24].

Folpet active principle studied by Canal-Raffin and collaborators [25] in the commercial formulation, Folpan 80WG®, showed a cytotoxic effect on human bronchial epithelial cells in vitro that could be in part explained by oxidative stress, lipid peroxidation and antioxidant activity.

The ecotoxicological approach based on biomarkers permits highlighting the sublethal effect caused by exposure to a specific contaminant or a mix of known or unknown contaminants. Biomarkers are a powerful tool for assessing the sub-lethal effects before irreversible damage occurs to organisms and colonies [26]. Alterations at a lower biological level can serve as important early warning signals to prevent macroscopic effects at the ecological level [26].

Our study aimed to investigate the toxicological effects on honeybees, used as model species, of the commercial formulation of two fungicides: FanticFNCWG® (folpet-based) and Ramedit combi® (cymoxanil + copper-based), alone and in combination. A multi-biomarker approach was applied to test neurotoxicity, measuring acetylcholinesterase (AChE) and carboxylcholinesterase (CaE) activities, metabolic alterations with alkaline phosphatase (ALP), the phase II of the detoxification process with glutathione S-transferase (GST), and the efficiency of the immune system by evaluating lysozyme (LYS) and phenoloxidase (proPO, PO) activities.

2. Materials and Methods

2.1. Chemicals

All the following chemicals were obtained from Sigma-Aldrich (St. Louis, MO, USA): monobasic and dibasic sodium phosphate, sodium chloride (NaCl), PBS, Tritons X-100, protease inhibitor cocktail powder; acetylthiocholine iodide (AcSCh.I), 5,5-dithio-bis(2, nitrobenzoic acid) (DTNB); 1-chloro-2,4-dinitrobenzene (CDNB), reduced L-glutathione (GSH); α-naphthyl acetate (αNA) and α-naphthyl butyrate (αNB); tris-hydroxy-methyl-aminomethane (Tris), magnesium chloride ($MgCl_2$), p-nitrophenyl phosphate (p-NPP); *Micrococcus lysodeikticus* solution, egg whites from chicken (HEL); monobasic potassium phosphate and bovine serum albumin (BSA), α-chymotrypsin, L-DOPA. BioRad Protein stain was obtained from BioRad (Segrate, Italy), while Diff-Quick dye was from Bio-optica (Milano, Italy). FANTIC F NCWG® was purchased from Gowan Italia S.r.l. (Faenza, Italy), and Ramedit combi® from Isagro S.p.A (Faenza, Italy).

2.2. Laboratory Exposure Condition

The laboratory experiments were carried out using worker honeybees about three days old that came from a synchronized brood from the same hive. During the experiment honeybees were fed with sucrose solution (50%) with the addition of treatments; the control consisted of only sucrose solution (50%). The feeding rate measured during the administration of the treatments and the mortality rate were checked daily. Honeybees were exposed to Fantic F NCWG® at 1 g/L (F1) and 2 g/L (F2), Ramedit combi® at 0.5 g/L (R0.5) and 1 g/L (R1), and the mix of the two (MIX1: F1 + R0.5 MIX2: F2 + R1) fungicides. The higher concentration used corresponded to the field application rates indicated on the label. The exposure was carried on for 6 days, following the previous protocol used [26]. To avoid a high mortality rate caused by captivity, the first day of the experiment was used for the acclimatation of bees. From the second day, each group of treated honeybees (25 bees in 2 cages for each treatment) was fed with a total of 12 mL of the treatment dissolved in sugar syrup, administered on alternate days.

The commercial fungicide Fantic F NCWG® composition was benalaxyl-m 3.75% and folpet 48.00% active ingredient and other coadjuvants not reported on the label, while Ramedit Combi® composition was cimoxanil 4.2% + metallic copper (in the form of copper oxychloride 39.75%) and other coadjuvants not on the label.

2.3. Samples Preparation and Enzyme Assays

At the end of the experiment, bees were anesthetized in ice (4 °C). The midgut was then removed with small forceps, and the head was separated from the rest of the body. All the samples were immediately frozen and stored at −80 °C. The brain was used to evaluate esterase activity (AChE and CaE). Metabolic and biotransformation enzymes (GST, ALP) and lysozyme activity (immune system) were evaluated on midgut extracts following the methodologies by Caliani and collaborators [26]. PO and proPO activities were evaluated in the thorax according to Burciaga and collaborators [27].

Tissue samples were pooled and weighed, and extraction medium was added in a volume corresponding to 10% (wt/vol) of the tissue. The buffer contained 40 mM Na phosphate buffer (pH 7.4), a mixture of protease inhibitors, and 1% Triton X-100. The samples were homogenized by a tissue lyser homogenizer (Qiagen, Hilden, Germany) for three periods of 30 s at 30 s intervals. After the homogenates were centrifuged at 4 °C for 20 min at 13,000× *g* for head and gut samples. The thorax and wings were removed, and together with the abdomen, they were homogenized in PBS and subsequently homogenized by a tissue lyser homogenizer (Qiagen) for three periods of 30 s at 30 s intervals. After the homogenization, the samples were centrifuged at 4 °C at 15,000× *g* for 15 min.

2.4. Pesticide Interaction Model

We applied the Pesticide Interaction Model according to Almasri et al. [28]. The IR (interaction ratio) considers the mortality rate of bees in the control group with respect to the treatment groups. The model obtained allows the identification of two different toxicological behaviors: the additive IR ≥ 1 or the synergistic IR > 1 effect. When the IR is <1, we have four other types of interaction: (i) purely antagonistic interaction when the mixture has a lower effect in comparison with the lowest toxic substance; (ii) sub-additive interaction, when the mixture has an effect higher than the most toxic substance but below the expected mortality; (iii) sub-additive synergistic effect when the toxicity of the mixture falls between the effects of the least toxic substance and the most toxic substance; or (iv) the effect of the mixture was considered independent when it caused a mortality rate similar to that of each pesticide individually [28].

2.5. Statistical Analysis

We tested significant differences in each biomarker between the control group, the fungicide FANTIC F NCWG®, the fungicide Ramedit Combi®, and the mixture samples using the Kruskal–Wallis (KW) non-parametric test. We applied Dunn's test with a Benjamini–Hochberg stepwise adjustment for pairwise multiple comparisons when the null hypothesis of the KW test was rejected. Statistical differences in food consumption were tested using Student's *t*-test. All analyses and graphical representations were implemented with RStudio (2023.06.09) and R (version 4.2.2).

3. Results

3.1. Laboratory Exposure

The feeding and the mortality rate are reported in Figure 1A,B. Sugar syrup intake was monitored after administering the treatment dose, with measurements taken for a total of two inputs. Mortality was recorded cumulatively after the experiment.

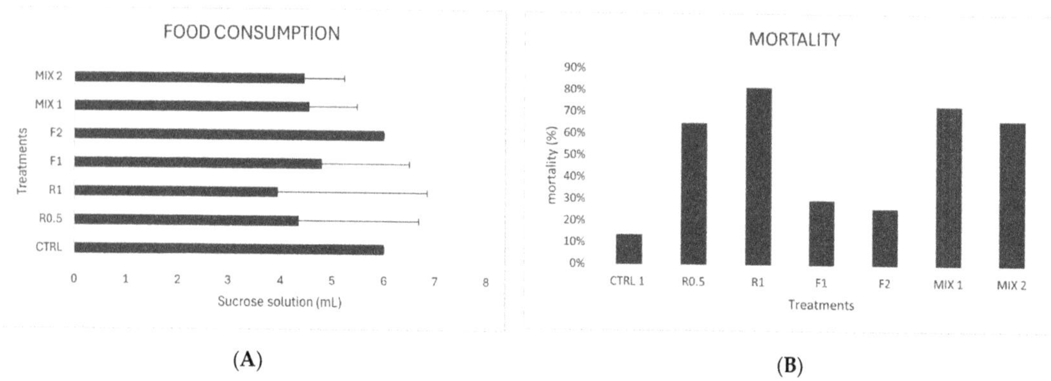

Figure 1. Feeding rate (mean and standard deviation (**A**)) and mortality (**B**), of the honeybees exposed to control (CTRL), fungicides Fantic®(F1, 1 g/L and F2, 2 g/L), Ramedit® (R0.5, 0.5 g/L and R1, 1 g/L), and two mixes of both fungicide (MIX1: F1 + R0.5; MIX2 F2 + R1).

All the treatments showed a reduction in food intake in the second input with respect to the control without a statistical difference between the inputs (*t*-test 1.97; $p = 0.096$). However, all the experimental groups had feeding rate values not lower than 66%. The feeding results confirm the full intake of the fungicides Fantic® and Ramedit® from the honeybees exposed.

The mortality rate of honeybee specimens exposed to the fungicides Fantic® and Ramedit® is presented in Figure 1. These data are necessary to apply the Pesticide Interaction Model. Following Almasri et al. [12], the higher toxicity of a compound is associated with higher mortality, so in order of toxicity: R1 (81%) > R0.5 > (65%) > F1 (30%) > F2 (26%). The results obtained show that concerning MIX1 and MIX2, we are in a sub-additive interaction of the third case.

3.2. Biomarkers Results

The results obtained from the evaluation of GST (Kruskal–Wallis (K-W): $X^2 = 15.3107$, $p = 0.02$) (Figure 2A) in specimens of *A. mellifera* exposed in the laboratory, showed higher values of activity after the exposure to Fantic® (F1), compared to the control group ($p = 0.0313$). On the other hand, a decrease in GST activity was observed in specimens exposed to the MIX2 compared to the bees exposed to F1 ($p = 0.0128$) and F2 ($p = 0.048$).

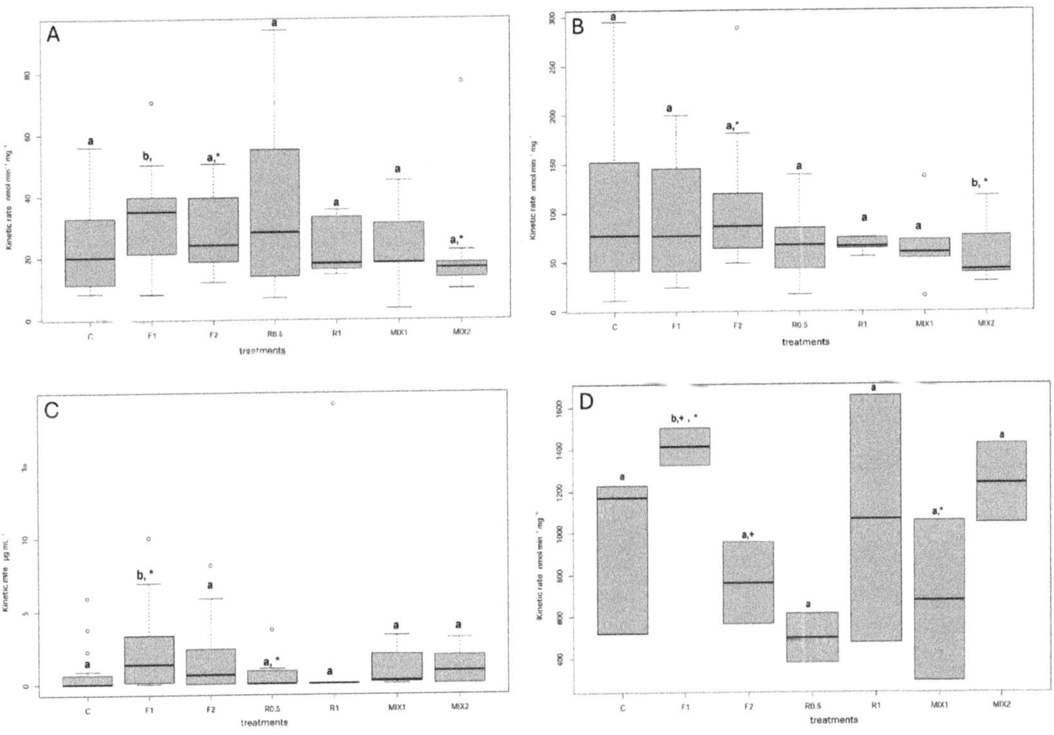

Figure 2. Activities of GST (**A**), ALP (**B**), LYS (**C**), and CaE (**D**) measured in honeybees (*Apis mellifera*) exposed to two different doses of Fantic® (F1; F2) and Ramedit® (R0.5; R1) and related mixtures (MIX1; MIX2). Different lowercase letters indicate significant differences with respect to control ($p < 0.05$); the same symbols ("*", "+") indicate significant differences between the group marked ($p < 0.05$).

The results of ALP activity (K-W: X^2: 16.7213, $p = 0.01$) (Figure 2B) showed lower values in MIX2 with respect to the control ($p < 0.024$) and between MIX2 and F2 ($p < 0.0497$) with lower values in MIX2.

The LYS activity (K-W: X^2: 19.8143, $p = 0.000$) (Figure 2C) showed an induction after all treatments compared to the control, except for treatment R1, which showed strong inhibitions in the activity. In particular, statistically significant differences were found between F1 and the control (p = 0.0017) and between F1 and R0.5 (p = 0.0058) and R1 (p = 0.0299) treatments.

The phenoloxidase pathway PO and proPO ratio showed higher values in the F1 (p = 0.041) and F2 (p = 0.0035) treatment with respect to the control.

The results of AChE activity evaluated in the brain tissues of honeybees exposed to the fungicides and related mixtures did not show any variation or neurotoxic effect with a median average activity of 2.5 (nmol min^{-1} mg $prot^{-1}$). Differently, the results of the CaE activity, evaluated in the same brain tissues of honeybees (Figure 2D), showed induction with a statistically significant difference between the F1 treatment and the control (p = 0.05). We also found a significant difference between the F1 and F2 (p = 0.0193) treatments, with a reduction in the activity from the lowest to the highest dose of the treatment. Moreover, we found statistically significant differences between F1 and MIX1 (p = 0.008) and between F1 and R0.5 (p = 0.0038). The CaE results for MIX1, composed of the sum of the F1 dose and R0.5 one, as well as for the GST data, confirmed the sub-additive effect in honeybees exposed to the mixtures.

4. Discussion

In our study we applied the Pesticide Interaction Model, proposed by Almasri et al. [28] and previously described, considering the mortality rate of bees to understand if the selected fungicides had additive, synergistic, or antagonistic effects on honeybees. The higher toxicity of a compound is associated with higher mortality, observing that R1 had the highest mortality rate, followed by R0.5, F1, and F2 (26%). The results obtained after the exposure to the mixtures of fungicides showed that we are in a sub-additive interaction of the third case. Almasri et al. [28] found the same interaction in *A. mellifera* exposed to the active ingredients of the fungicide difenoconazole (F), the insecticide imidacloprid (I), and the herbicide glyphosate (H) for 20 days.

The glutathione S-transferase is a class of enzymes whose primary role is to facilitate the conjugation of the antioxidant molecule glutathione (GSH) with various electrophilic and hydrophobic compounds, such as toxic substances like pesticides and reactive oxygen species (ROS). The increase in GST activity was observed by Caliani et al. [26] in specimens of honeybees topically exposed to concentrations of 100 g/L and 200 g/L of the fungicide AmistarXtra®. In the study by Han et al. [29] conducted on *Apis cerana*, the GST activity of newly emerged bees was reduced following exposure to the fungicide propiconazole, while no statistically significant difference in GST activity was found between the exposed groups and the control group for forager bees. The results obtained in the present work suggest that exposure to the fungicides Fantic® and Ramedit® may influence the detoxification activity of GST. It can both increase the activity compared to the control bees and inhibit the activity when mixed at a lower concentration.

Glutathione and other thiol-containing molecules play an essential role in the inactivation of the folpet active principle of Fantic®. It has been described that the presence of cytoplasmatic GSH reduces the folpet concentration in the tissues of several species [30]. Also, the reduction in GSH availability induced the activity of GST to regenerate GSH. Our results could highlight this mechanism in the exposure to Folpet-based fungicide.

Alkaline phosphatase is a class of enzymes that helps break down phosphate-containing nutrients vital for the insect's growth and energy, such as nucleotides and phospholipids, into simpler forms that can be absorbed. It is also involved in the immune response, particularly in the gut. It can participate in the detoxifying process of toxins and assist in protecting the gut lining against harmful substances like pathogens. The ALP inhibition in honeybees exposed to the MIX1 and MIX2 compared to other treatments was probably due to the presence of copper in the Ramedit® commercial fungicide. Since there was a statistical difference between F2 and MIX2, we assumed that Folpet was not responsible for the inhibition in the mix. Our study was the first study to investigate the toxicological effects on alkaline phosphatase of copper-based fungicides and cymoxanil. Until now, other studies have examined the effects on the ALP enzyme for different types of metals and fungicides: Vlahovic et al. [31] highlighted that the ALP activity was inhibited following exposure to cadmium. The study by Badiou-Beneteau et al. [32] on bees exposed to lower doses (2.56 ng bee^{-1}) of the insecticide thiamethoxam showed a 20% increase in ALP activity compared to the control. Inhibition of ALP was also observed in the study by Caliani et al. [26] following exposure of *A. mellifera* specimens to concentrations of 100 g/L and 200 g/L of the fungicide AmistarXtra, concentrations of 0.1 and 2.5 g/L of cadmium, and a concentration of 12.4 g/L of EMS.

Lysozyme is synthesized in the fat body of honeybees [33] and plays an essential role in the innate immune system of insects, aiding in the defense against microbial infections, especially bacteria. Insects depend heavily on their innate immune responses, as they lack the adaptive immune systems found in vertebrates. Lysozyme is a useful biomarker for investigating disease resistance and therefore the immune system status of an organism. The study by Caliani et al. [26] exposed *A. mellifera* to the fungicide AmistarXtra (100 g/L and 200 g/L), cadmium (0.1 g/L and 2.5 g/L), and EMS (12.4 g/L), resulting in a statistically significant decrease in LYS activity compared to the control after both cadmium doses, the highest fungicide dose, and the EMS dose. These results were in line with the findings of our work, where bees exposed to copper-based fungicide caused a reduction in the immunological capacity of lysozyme. Differently, the folpet-based fungicides caused an induction of the lysozyme activity. Furthermore, LYS activity is negatively correlated with ALP activity in the experimental groups R0.5 ($\rho = -0.46$; $p < 0.05$) and F1 ($\rho = -0.42$, $p < 0.05$). As the activity of ALP increases, the activity of the immune system decreases, indicating that the detoxifying enzyme is working while the immune system is being taxed. LYS is also correlated with GST in the experimental groups exposed to the lowest ($\rho = -0.52$; $p < 0.05$) and the highest concentration ($\rho = -0.44$, $p < 0.05$) of Ramedit®, specifically indicating that as GST detoxifying activity increases, the functionality of the immune system decreases due to the presence of toxic substances in the honeybee's body. Therefore, our results suggest that exposure to the fungicides Fantic® and Ramedit® causes an alteration in the immune system of bees, increasing lysozyme activity, especially following exposure to Fantic® and to the mixtures of the two fungicides.

Another critical component of the immune system in invertebrates is the phenoloxidase pathway. It plays an essential role in insect defense against pathogens such as bacteria, fungi, and parasites [34]. The central enzyme in this pathway is prophenoloxidase (proPO), a zymogen of phenoloxidase (PO). Upon exposure to pathogens or injury, the proPO is activated into phenoloxidase (PO) by proteolytic cleavage [35]. This activation is usually mediated by serine proteases that are part of the immune response. Phenoloxidase catalyzes the oxidation of phenols to quinones that subsequently polymerize to form melanin, which is deposited around the invading pathogen [36]. The PO/proPO ratio indicates that efficiency increases as the value approaches one. The preliminary laboratory results of this biomarker obtained in this work showed the activation of PO in the specimens exposed to both the Fantic treatments. Similarly to lysozyme, exposure to Fantic triggers an increase in phenoloxidase, activating key components of the immune system.

The esterase enzymes, acetylcholinesterase and carboxylesterase, play key roles in the transmission of nerve impulses, with acetylcholinesterase being particularly crucial for this process. Carboxylesterase is a class of enzymes which are also involved in the detoxification of neurotoxic substances. The absence of neurotoxicity in all the treatments highlighted by the AChE results could be related to the induction of CaE, which plays a role in phase I detoxification processes and also serves a defensive function by protecting AChE from inactivation [37].

5. Conclusions

Laboratory experiments have demonstrated that the commercial formulation of fungicides, containing multiple active ingredients, produces different effects compared to exposure to individual molecules. By applying interaction models and conducting a multi-biomarker evaluation, we were able to more accurately explain the interactions between compounds when mixed. This approach allowed us to better replicate what occurs under environmentally realistic conditions. Biomarker responses revealed that these complex mixtures impact various enzymatic pathways, such as the immune system, altering fundamental enzyme kinetics involved in the detoxification process and potentially disrupting normal bee functions. This study underscores the need for a more thorough investigation of the sublethal effects of commercial pesticides, especially those used in vineyards, which are less studied compared to compounds applied in orchards. Such research is crucial for the protection of managed and wild pollinators, expanding our understanding of pesticide impacts and their role in the Colony Collapse Disorder (CCD) phenomenon.

Author Contributions: Conceptualization, T.C. and S.C.; methodology, T.C.; formal analysis, T.C.; investigation, T.C. and G.M.; resources, S.C.; data curation, T.C.; writing—original draft preparation, T.C.; writing—review and editing, A.D.N., I.C. and S.C.; supervision, I.C. and S.C.; funding acquisition, S.C. All authors have read and agreed to the published version of the manuscript.

Funding: This research was funded by FSE-REACT EU for the PON RICERCA E INNOVAZIONE 2014–2020, action IV.6 GREEN: Counteracting the decline of pollinators: Evaluating the environmental sustainability of pesticides and developing bee-friendly agroecological strategies to improve pollinator health.

Institutional Review Board Statement: Not applicable.

Informed Consent Statement: Not applicable.

Data Availability Statement: The raw data supporting the conclusions of this article will be made available by the authors on request.

Conflicts of Interest: The authors declare no conflicts of interest.

References

1. Yang, Y.; Wu, Y.; Long, H.; Ma, X.; Shariati, K.; Webb, J.; Guo, L.; Pan, Y.; Ma, M.; Deng, C.; et al. Global honeybee health decline factors and potential conservation techniques. *Food Secur.* **2023**, *15*, 855–875. [CrossRef]
2. Hristov, P.; Shumkova, R.; Palova, N.; Neov, B. Factors associated with honey bee colony losses: A mini-review. *Vet. Sci.* **2020**, *7*, 166. [CrossRef] [PubMed]
3. Harwood, G.P.; Dolezal, A.G. Pesticide–virus interactions in honey bees: Challenges and opportunities for understanding drivers of bee declines. *Viruses* **2020**, *12*, 566. [CrossRef] [PubMed]
4. Monchanin, C.; Burden, C.; Barron, A.B.; Smith, B.H. Heavy metal pollutants: The hidden pervasive threat to honey bees and other pollinators. In *Advances in Insect Physiology*; Elsevier: Amsterdam, The Netherlands, 2023; pp. 255–288.
5. Tudi, M.; Daniel Ruan, H.; Wang, L.; Lyu, J.; Sadler, R.; Connell, D.; Chu, C.; Phung, T.D. Agriculture development, pesticide application and its impact on the environment. *Int. J. Environ. Res. Public Health* **2021**, *18*, 1112. [CrossRef] [PubMed]
6. Agata, D.N.; Silvia, C.; Tommaso, C.; Giampiero, C.; Ilaria, C. Review on Sublethal Effects of Environmental Contaminants in Honey Bees (*Apis mellifera*), Knowledge Gaps and Future Perspectives. *Int. J. Environ. Res. Public Health* **2021**, *18*, 1863. Available online: https://www.mdpi.com/1660-4601/18/4/1863 (accessed on 18 September 2024). [CrossRef]
7. Martinello, M.; Manzinello, C.; Borin, A.; Avram, L.E.; Dainese, N.; Giuliato, I.; Gallina, A.; Mutinelli, F. A Survey from 2015 to 2019 to Investigate the Occurrence of Pesticide Residues in Dead Honeybees and Other Matrices Related to Honeybee Mortality Incidents in Italy. *Diversity* **2020**, *12*, 15. [CrossRef]

8. Ye, H.; Jiang, J.; Lei, Y.; Fang, N.; Luo, Y.; Cheng, Y.; Li, Y.; Wang, X.; He, H.; Yu, J.; et al. A systemic study of cyenopyrafen in strawberry cultivation system: Efficacy, residue behavior, and impact on honeybees (*Apis mellifera* L.). *Environ. Pollut.* **2024**, *346*, 123601. [CrossRef]
9. Puig-Montserrat, X.; Stefanescu, C.; Torre, I.; Palet, J.; Fàbregas, E.; Dantart, J.; Arrizabalaga, A.; Flaquer, C. Effects of organic and conventional crop management on vineyard biodiversity. *Agric. Ecosyst. Environ.* **2017**, *243*, 19–26. [CrossRef]
10. Bleyer, G.; Schumacher, S.; Fuchs, R. Sulfur—A potential additive to increase the efficacy of copper-based fungicides against grapevine downy mildew. *OENO One* **2024**, *58*. [CrossRef]
11. Viret, O.; Gindro, K. Disease Control. In *Science of Fungi in Grapevine*; Springer: Berlin/Heidelberg, Germany, 2024; pp. 401–461.
12. Romanazzi, G.; Mancini, V.; Feliziani, E.; Servili, A.; Endeshaw, S.T.; Neri, D. Impact of alternative fungicides on grape downy mildew control and vine growth and development. *Plant Dis.* **2016**, *100*, 739–748. [CrossRef]
13. Lamichhane, J.R.; Osdaghi, E.; Behlau, F.; Köhl, J.; Jones, J.B.; Aubertot, J.N. Thirteen decades of antimicrobial copper compounds applied in agriculture. A review. *Agron. Sustain. Dev.* **2018**, *38*, 28. [CrossRef]
14. Rusjan, D.; Strlič, M.; Pucko, D.; Korošec-Koruza, Z. Copper accumulation regarding the soil characteristics in Sub-Mediterranean vineyards of Slovenia. *Geoderma* **2007**, *141*, 111–118. [CrossRef]
15. *Commission Regulation (EC) No. 473/2002*; European Commission: Bruxelles, Belgium, 2002.
16. European Commission. *Commission Implementing Regulation (EU) No 540/2011 of 25 May 2011 Implementing Regulation (EC) No 1107/2009 of the European Parliament and of the Council as Regards the List of Approved Active Substances (Text with EEA Relevance) Text with EEA Relevance*; European Commission: Bruxelles, Belgium, 2023.
17. Russell, P. A century of fungicide evolution. *J. Agric. Sci.* **2005**, *143*, 11–25. [CrossRef]
18. Thind, T. Fungicides in Plant Disease Management. *Approaches Trends Plant Dis. Manag.* **2014**, *13*, 243–263.
19. Ahmed, M.S.; Massoud, A.H.; Derbalah, A.S.; Al-Brakati, A.; Al-Abdawani, M.A.; Eltahir, H.A.; Yanai, T.; Elmahallawy, E.K. Biochemical and histopathological alterations in different tissues of rats due to repeated oral dose toxicity of cymoxanil. *Animals* **2020**, *10*, 2205. [CrossRef]
20. Cheng, B.; Zhang, H.; Hu, J.; Peng, Y.; Yang, J.; Liao, X.; Liu, F.; Guo, J.; Hu, C.; Lu, H. The immunotoxicity and neurobehavioral toxicity of zebrafish induced by famoxadone-cymoxanil. *Chemosphere* **2020**, *247*, 125870. [CrossRef] [PubMed]
21. Gupta, P. Herbicides and fungicides. In *Biomarkers in Toxicology*; Elsevier: Amsterdam, The Netherlands, 2019; pp. 477–499.
22. Christen, V.; Krebs, J.; Fent, K. Fungicides chlorothanolin, azoxystrobin and folpet induce transcriptional alterations in genes encoding enzymes involved in oxidative phosphorylation and metabolism in honey bees (*Apis mellifera*) at sublethal concentrations. *J. Hazard. Mater.* **2019**, *377*, 215–226. [CrossRef]
23. Ham, J.; Yun, B.H.; Lim, W.; Song, G. Folpet induces mitochondrial dysfunction and ROS-mediated apoptosis in mouse Sertoli cells. *Pestic. Biochem. Physiol.* **2021**, *177*, 104903. [CrossRef]
24. Leeb, C.; Schuler, L.; Brühl, C.A.; Theissinger, K. Low temperatures lead to higher toxicity of the fungicide folpet to larval stages of Rana temporaria and Bufotes viridis. *PLoS ONE* **2022**, *17*, e0258631. [CrossRef] [PubMed]
25. Canal-Raffin, M.; L'azou, B.; Jorly, J.; Hurtier, A.; Cambar, J.; Brochard, P. Cytotoxicity of folpet fungicide on human bronchial epithelial cells. *Toxicology* **2008**, *249*, 160–166. [CrossRef]
26. Caliani, I.; Campani, T.; Conti, B.; Cosci, F.; Bedini, S.; D'agostino, A.; Ammendola, A.; Di Noi, A.; Gori, A.; Casini, S. Multi-biomarker approach and IBR index to evaluate the effects of different contaminants on the ecotoxicological status of *Apis mellifera*. *Ecotoxicol. Environ. Saf.* **2021**, *208*, 111486. Available online: https://www.scopus.com/inward/record.uri?eid=2-s2.0-85094624286&doi=10.1016/j.ecoenv.2020.111486&partnerID=40&md5=a3cc12e4c4b8014a23035794cdb12065 (accessed on 22 October 2023). [CrossRef]
27. Burciaga, R.A.; Ruiz-Guzmán, G.; Lanz-Mendoza, H.; Krams, I.; Contreras-Garduño, J. The honey bees immune memory. *Dev. Comp. Immunol.* **2023**, *138*, 104528. [CrossRef]
28. Almasri, H.; Tavares, D.A.; Pioz, M.; Sené, D.; Tchamitchian, S.; Cousin, M.; Brunet, J.-L.; Belzunces, L.P. Mixtures of an insecticide, a fungicide and a herbicide induce high toxicities and systemic physiological disturbances in winter Apis mellifera honey bees. *Ecotoxicol. Environ. Saf.* **2020**, *203*, 111013. [CrossRef]
29. Han, W.; Yang, Y.; Gao, J.; Zhao, D.; Ren, C.; Wang, S.; Zhao, S.; Zhong, Y. Chronic toxicity and biochemical response of Apis cerana cerana (Hymenoptera: Apidae) exposed to acetamiprid and propiconazole alone or combined. *Ecotoxicology* **2019**, *28*, 399–411. [CrossRef] [PubMed]
30. Arce, G.T.; Gordon, E.B.; Cohen, S.M.; Singh, P. Genetic toxicology of folpet and captan. Critical Reviews in Toxicology. *Crit. Rev. Toxicol.* **2010**, *40*, 546–574. [CrossRef] [PubMed]
31. Vlahović, M.; Lazarević, J.; Perić-Mataruga, V.; Ilijin, L.; Mrdaković, M. Plastic responses of larval mass and alkaline phosphatase to cadmium in the gypsy moth larvae. *Ecotoxicol. Environ. Saf.* **2009**, *72*, 1148–1155. [CrossRef]
32. Badiou-Bénéteau, A.; Carvalho, S.M.; Brunet, J.-L.; Carvalho, G.A.; Buleté, A.; Giroud, B.; Belzunces, L.P. Development of biomarkers of exposure to xenobiotics in the honey bee Apis mellifera: Application to the systemic insecticide thiamethoxam. *Ecotoxicol. Environ. Saf.* **2012**, *82*, 22–31. [CrossRef]

33. Yazlovytska, L.S.; Karavan, V.V.; Domaciuk, M.; Panchuk, I.I.; Borsuk, G.; Volkov, R.A. Increased survival of honey bees consuming pollen and beebread is associated with elevated biomarkers of oxidative stress. *Front. Ecol. Evol.* **2023**, *11*, 1098350. Available online: https://www.frontiersin.org/journals/ecology-and-evolution/articles/10.3389/fevo.2023.1098350 (accessed on 6 November 2023). [CrossRef]
34. Binggeli, O.; Neyen, C.; Poidevin, M.; Lemaitre, B. Prophenoloxidase activation is required for survival to microbial infections in Drosophila. *PLoS Pathog.* **2014**, *10*, e1004067. [CrossRef] [PubMed]
35. Cerenius, L.; Söderhäll, K. Immune properties of invertebrate phenoloxidases. *Dev. Comp. Immunol.* **2021**, *122*, 104098. [CrossRef] [PubMed]
36. Liu, H.; Jiravanichpaisal, P.; Cerenius, L.; Lee, B.L.; Söderhäll, I.; Söderhäll, K. Phenoloxidase is an important component of the defense against Aeromonas hydrophila infection in a crustacean, Pacifastacus leniusculus. *J. Biol. Chem.* **2007**, *282*, 33593–33598. [CrossRef]
37. Jackson, C.J.; Liu, J.W.; Carr, P.D.; Younus, F.; Coppin, C.; Meirelles, T.; Lethier, M.; Pandey, G.; Ollis, D.L.; Russell, R.J.; et al. Structure and function of an insect α-carboxylesterase (α Esterase 7) associated with insecticide resistance. *Proc. Natl. Acad. Sci. USA* **2013**, *110*, 10177–10182. [CrossRef] [PubMed]

Journal of
Xenobiotics

Review

Developments in Toxicity Testing with Duckweeds

Paul Ziegler

Department of Plant Physiology, University of Bayreuth, 95440 Bayreuth, Germany; paul.ziegler@uni-bayreuth.de

Abstract: Duckweeds are a family of small floating macrophytes (the Lemnaceae) that inhabit quiet freshwaters worldwide. They have long been employed to determine toxicity to higher plants in the aquatic environment, and standardized national and international protocols have been developed for this purpose using two representative species. While these protocols, which assess the growth of the leaf-like fronds of the tested duckweed, are indeed suitable and still frequently used for detecting the toxicity of water-borne substances to aquatic higher plant life, they are cumbersome and lengthy, determine endpoints rather than depict toxicity timelines, and provide no information as to the mechanisms involved in the indicated toxicity. Progress has been made in downscaling, shortening and improving the standardized assay procedures, and the use of alternative duckweed species, protocols and endpoints for detecting toxicity has been explored. Biomarkers of toxic effect have long been determined concomitantly with testing for toxicity itself, and their potential for the assessment of toxicity has recently been greatly expanded by transcriptomic, proteomic and metabolomic techniques complemented by FITR spectroscopy, transformation and genotoxicity and timescale toxicity testing. Improved modern biomarker analysis can help to both better understand the mechanisms underlying toxicity and facilitate the identification of unknown toxins.

Keywords: duckweeds; toxicity tests; biomarkers

1. Introduction

The influx of industrial, agricultural, municipal and domestic waste into natural waters poses serious hazards for life forms inhabiting or consuming these waters, including bacteria, plants, invertebrates, fish and mammals, including man [1,2]. The study of the effects of water pollutants on aquatic plant life is particularly important from the viewpoint of plants being primary producers and thus supportive of other aquatic organisms along the food chain and that they also pass on toxic substances that they may have taken up to organisms consuming them. Aquatic plant life includes micro- and macroalgae, bryophytes, pteridophytes and flowering plants, species of each of which are used in testing for deleterious effects of water pollutants [2]. Microalgae have long been the most frequently used life form for investigating the effects of water contaminants on aquatic plant life (see [2,3]). However, flowering plants come next in this regard [2], most prominently rooted species of the genus *Myriophyllum* (water milfoils) and duckweeds [2], a widespread family of small, simply constructed floating macrophytes (the Lemnaceae) that inhabit quiet or slow-flowing bodies of freshwater [4–6]. Even though phytotoxicity studies on aquatic plants have traditionally been of minor importance in establishing regulatory measures concerning the environmental hazard of water contaminants (see [2,3]), duckweeds have a long history of being used to determine the toxic potential of water contaminants for aquatic higher plant life and to understand the toxicities involved [7]. The resurgence of interest in duckweeds as versatile model plants [8] helps consolidate the continuing importance of these macrophytes in the field of aquatic phytotoxicity.

Duckweeds are monocots that consist of flattened or rounded leaf- and thallus-like assimilatory organs or fronds that range from 1 to 15 mm in diameter or length and are only a few cells in thickness [5,8]. The fronds consist largely of spongy mesophyll with extensive air-filled intercellular spaces that confer buoyancy and some bear hairless adventitious root on the underside [5,8]. There are 35 species of duckweed [9,10] distributed among the five genera *Spirodela* (abbreviation *S.*), *Landoltia* (*La.*), *Lemna* (*Le.*), *Wolffiella* (*Wa.*) and *Wolffia* (*Wo.*), in addition to two interspecies hybrids [11–14] that exhibit genus-typical differences in the size and complexity of the fronds and in the number of roots they bear [5,6,8,15]; for pictorial representations, see [8,16,17]. Although they can flower and reproduce sexually, duckweeds grow mainly by vegetative propagation, with daughter fronds budding off from one or two meristematic pouches or pockets in the mother fronds [5,8]. This can take place very rapidly: duckweeds constitute the most rapidly growing higher plants [18,19]. The fronds remain temporarily attached to one another via connecting stalks or stipes, forming colonies of interconnected fronds that spread out over the water surface to ensure optimal access to water nutrients [8,20–22] that are taken up via the ventral frond surfaces and the roots (see [8]).

The rapid growth potential of duckweeds and their ability to take up numerous water-contaminating compounds in addition to nutrients have made these macrophytes promising agents for the remediation of industrial, agricultural and domestic wastewaters [8,16,17]. These attributes—combined with small size, simple structure and a consequent ease of cultivation and analysis—also facilitate the detection and interpretation of detrimental effects of water contaminants on aquatic higher plant material. Duckweeds accordingly are well-suited as model organisms for aquatic phytotoxicity testing in their amenability to the collection of data relevant to a variety of research questions, backed up by well-developed experimental protocols, a comprehensive literature base and an international expert community (see [23]). The International Steering Committee on Duckweed Research and Applications (ISCDRA) helps to guide, coordinate and communicate research on duckweeds, as detailed in the quarterly released Duckweed Forum newsletter that can be accessed via the website of the Rutgers Duckweed Stock Cooperative (RDSC; ruduckweed.org: see [8]).

2. Standardized Duckweed Toxicity Testing

Duckweeds have long been employed for testing for toxic effects of water-borne substances—often typical water contaminants—on higher plant life in aqueous environments: the pertinent literature up till 1990 was reviewed by Wang [7]. Based on these experiences, several national and international organisations concerned with the aquatic environment set up protocols for testing for water-borne toxicity to aqueous higher plants using either *Lemna minor* or *Lemna gibba* as model test organisms [16]: see Table 1.

These "classical" tests exemplified by the OECD Test No. 221 [24], the ISO test 20079 [25] and the USEPA test [26] summarized in [27] consist of the incubation of 9–12 fronds of a duckweed (*Le. minor* or *Le. gibba*: see [1] for photographic images) in at least 100 mL of mineral salt medium (SIS, 20xAAP or Steinberg) in correspondingly large vessels, such as beakers or culture dishes, both without and with several concentrations of a putatively toxic substance. After 7 days, growth parameters, including frond number, area and weight, are measured, and the chlorophyll content of the fronds can also be assessed. A more recent equivalent procedure has been prescribed by the Chinese agricultural industry standard NY/T 3090–2007 [28] (see [29]). An inhibition of frond growth and a reduction in the chlorophyll content of the cultures containing the test substance in comparison to the control define the toxicity of the test substance to the duckweed. Toxicity parameters can be calculated from dose-response curves for effective doses or concentrations (ECs) of test substances at effect levels such as 50%, 20% or 10% [30], e.g., EC_{50}.

Table 1. National and international organisations that have developed standardized tests for water-borne toxicity to aquatic higher plants with duckweeds from 1990 to 2006. Bibliographic information for the standardized tests can be found in [27,31].

Organisation	Year	Duckweed	Country/Area
ITM: Institute of Applied Environmental Research	1990	*Lemna minor*	Sweden
ASTM: American Society for Testing and Materials	1991	*Lemna gibba*	USA
APHA: American Public Health Association	1995	*Lemna gibba*	USA
SIS: Swedish Standards Institute	1995	*Lemna minor*	Sweden
AFNOR: Association Française de Normalisation	1996	*Lemna minor*	France
USEPA: US Environmental Protection Agency	1996	*Lemna minor/gibba*	USA
EC: Environment Canada	1999	*Lemna minor*	Canada
ISO: International Organisation for Standardization	2005	*Lemna minor*	International
OECD: Organisation for Economic Cooperation and Development	2006	*Lemna minor/gibba*	International

Duckweed growth proceeds exponentially over time under non-limiting growth conditions, since each frond replicates regularly. In this case, there is a linear relationship between frond number, area or weight (F), and time (t): the slope of the line represents the relative growth rate (RGR), which can be described as follows (see [18]):

$$RGR = \ln F_t - \ln x F_{t0}/t - t0$$

It has been recommended that the doubling time of frond number in the control should be less than 2.5 days throughout the 7-day test period, which corresponds to an approximately 7-fold increase in frond number and a RGR of 0.275 d^{-1} [24], a criterion easily met by representative *Le. minor* and *Le. gibba* clones [18]. The RGR must remain constant throughout the incubation time of the assay to assure that assay conditions are not limiting the intrinsic growth rate of the duckweed. This means that—for the reliable assessment of phytotoxicity in terms of growth inhibition—the volume of medium, the nutrient content of the medium and the inoculum size must be designed to ensure that exponential growth can occur throughout the test period [32]. If the mineral salt content of the nutrient medium, for example, is not sufficient to maintain exponential growth of the control fronds throughout the assay duration, it might be sufficient to enable growth slowed by toxic influences to proceed in a continual exponential fashion that could thus not accurately be related to the control. The preliminary work required to ensure exponential control growth was important in designing the standard toxicity tests. According to the

OECD, the necessary criteria for the validity of these assays can be attained using a static test regime, in which the test conditions at the start of the test are left without interference until the growth parameter measurements are made at the end of the 7-day incubation period [24], and this regime is usually adhered to.

An important factor for the comparability of toxicity data obtained with a standardized duckweed toxicity test over extended periods of time, including seasonal changes, is the stability of the intrinsic growth of the test organism. This stability ensures the re-usability of previously determined toxicity data in the context of new investigations. In a corresponding study, control growth data for toxicity tests with the same *S. polyrhiza* clone over a period of four years under identical test conditions revealed a good long-term stability of frond number and area parameters, although some variability with time and the season was noted [33].

The standardized tests outlined above are indeed well-suited for determining and defining overall toxicity to a duckweed because growth inhibition is indicative of global disturbances in the functioning of the test organism. This also applies to reduced chlorophyll synthesis—including the bleaching that often accompanies this—that reflects the loss of the photosynthetic potential that is the signature characteristic of most plant life. It is emphasized that the standardized tests are "toxicity tests" in the narrow sense of being concerned only with determining the deleterious potential *per se* of questionable substances on the test organism under defined, reproducible conditions in the laboratory. This sets them apart somewhat from "ecotoxicity testing", which investigates the deleterious effects of water contaminants on duckweeds under conditions that are ecologically relevant, i.e., with respect to contaminant concentrations, the constitution of the medium and the surrounding conditions, and the duration and frequency of exposure [2]. Ecotoxicological testing with duckweeds makes use of the machinery of the "toxicity testing" discussed here to investigate the actual consequences of water pollution on aquatic higher plant life.

The standardized tests are relatively easy to carry out but are cumbersome due to the use of large culture vessels and culture medium volumes and are lengthy. Efforts to improve and complement the classical testing procedures are discussed in Section **3**. The standardized tests measure only overall growth-related endpoints and provide no information as to the course or the mechanism of the toxicity. However, the investigation of toxicity-related biomarkers for mechanistic interpretation of toxic effects has a long history of having been carried out parallel to the determination of toxicity itself. Section **4** deals with developments to improve the amount and quality of biomarker information obtainable for better mechanistic understanding of overall toxicity.

3. Determining Toxicity

One approach toward improving and extending the "classical" determination of the toxicity of water contaminants to duckweeds as laid down by the standardized toxicity tests has been largely concerned with efforts to miniaturize and speed up the standardized assay procedures with an eye to convenience and efficiency in carrying out large numbers of toxicity tests. This is discussed in the following Section 3.1. On the other hand, the possibility of using duckweed species other than *Le. minor* and *Le. gibba* as test organisms and the assessment of novel toxicity endpoints to good advantage has also been examined. These "alternative" developments are discussed in Section 3.2.

3.1. Modifying the Standardized Tests

Issues for improving the techniques of standardized duckweed assays proposed by the organizations shown in Table 1 for detecting toxicity have been the scaling down the of the assay dimensions, the shortening of the incubation period required to assess toxicity, ensuring the quality of the incubation medium during the duration of the test and the development of innovative procedures for growth parameter measurement and analysis.

3.1.1. Downsizing and Timesaving

The standardized duckweed toxicity test assays recommend the use of relatively large culture vessels, medium volumes and duckweed innocula. The downsizing, or miniaturization, of the standard test format offers the advantages of reduced space requirements, lower costs, reduction in test solution volumes, less waste disposal and less starting plant material. Experimentation in this regard had already been carried out prior to the development of the standardized tests. Two reports from 1990 described a reduction in the solution volume for toxicity testing with *Le. minor* to 15 mL using 60 × 15 mm culture dishes and plastic cups [7,34]. Downsized versions of the standardized tests were later proposed in which five to six fronds of *Le. minor* are incubated in 10 mL of culture medium in six-well culture plates for 7 days [35,36]. The results obtained with tests of the toxicity of KCl and 3,5-dichlorophenol and of Zn nanoparticles in these miniaturized systems were reported to be comparable to those obtained with full-size standardized tests. A recent example of the use of this test format is the investigation of the toxicity of four antidepressants to *Le. minor* [37].

Further miniaturization had been reported as early as 2004, where two three-frond colonies of *Lemna paucicostata* (a synonym for *Lemna aequinoctialis*) were incubated for 7 days in 5 mL portions of medium containing numerous individual herbicides with different modes of action in 6-well culture plates [38], and two-frond colonies of *Le. gibba* were incubated for 7 days in 2.5 mL of test solution containing olive mill waste in the wells of 12-cell culture plates in 2007 [39]. Miniaturization has recently been described for the investigation of seven to eight fronds of *Le. minor* in 2 mL of medium containing various individual pesticides in the wells of 24-well culture plates [40]. It is questionable whether toxicity values determined with the formats of the two latter investigations indeed reflect the values that would be obtained with standardized test formats. The much higher frond number/medium volume ratios used in these studies in comparison with the standardized test formats may readily lead to frond crowding and nutrient limitation, which can quite generally be an issue in miniaturization. This would make the maintenance of exponential duckweed growth and a continuous RGR over the 7-day incubation period unlikely, which would—as discussed above—compromise the accuracy and reliability of the toxicity test. This has been indicated in an investigation in which the incubation of two fronds in 2 mL medium portions led to slower frond growth and deviant toxicity values in comparison with the results obtained in parallel tests with the standardized format [36] (see also the last paragraph of this section).

The recommended incubation period of the test substance with the duckweeds in the standardized assay procedures is 7 days. Although this exposure duration may actually be too short in terms of ecotoxicological testing [2], a shortening of the incubation period with the test substance can be advantageous for testing the toxicity of large numbers of samples in screening procedures and can also reduce the tendencies of frond crowding and nutrient limitation. Incubation times shorter than 7 days had already been employed prior to the establishment of the standardized tests (4 d in the case of *Le. minor* in 1986 (see [7]) but have seldom been used in the standardized formats since then. Two relatively recent examples are 4-day incubation periods in test procedures with *Landoltia punctata* in the presence of nanoparticular and soluble Cu [41] and with *Le. minor* and *Le. gibba* exposed to the pharmaceuticals chlorpromazine, paracetamol and diclofenac [42].

The miniaturization and incubation time reduction of the standardized assays can be combined. One recent example is the assessment of the phytotoxic effects of several environmentally relevant metals and metalloids to three to four fronds of *Le. gibba* incubated for three days in a test volume of 4 mL in 12-well culture plates by means of chlorophyll fluorescence imaging [43]. Growth and photosynthetic parameters were measured in the same samples. The inhibition of relative growth rates in terms of frond number and area were calculated from the obtained chlorophyll fluorescence images. Such multi-well, plate-based duckweed phytotoxicity assays significantly reduce the space, time and sample volume requirements of the standard duckweed growth inhibition tests. These benefits may, however, be compromised by lowered test sensitivity, with considerably higher EC_{50} values for growth inhibition being recorded in comparison with data obtained for standardized tests [43]. Nevertheless, multi-well plate-based duckweed phytotoxicity assays offer definite advantages in the rapid screening of large sample numbers. The benefits of miniaturization and short incubation times have become still more impressive, however, in the development of the duckweed phytotoxicity assays making use of alternative toxicity indictors to be discussed in Section 3.2.2.

3.1.2. Ensuring the Quality of the Incubation Medium

Most toxicity tests with duckweeds proceed with incubation medium for duckweed growth containing the test substance being provided at the initiation of the test and not interfered with thereafter. The nutrients of the incubation medium in such "static" standard assays (already referred to in Section 2) can become limiting during the 7-day test duration when the ratio of the inoculum to the incubation volume is too high and the incubation period is too long, especially in controls exhibiting vigorous growth. In addition, the concentration of test substances in the medium can also decrease significantly over long incubation periods due to take-up by the duckweed and especially when the test substances are unstable or volatile (see [44]). The initial composition of the medium (including the designated test substance concentration) can be approximately maintained when the incubation medium is periodically renewed and can be kept quite uniform in a flow-through setup. Protocols for employing medium renewal and flow-through in testing with *Le. minor* and *Le. gibba* have long been available (e.g., [45,46]), and the standardized assays allow for static, renewal and flow-through experimental formats (see, e.g., [24,33]).

A protocol based on the static ASTM test (see Table 1) for maintaining the quality of the medium in testing with *Le. gibba* was developed, according to which the medium containing the test substance is renewed daily [44]. This daily renewal ensures an approximately constant nutrient supply and test substance concentration and thus exposure to the test substance throughout the 7-day test period. However, such test procedures are considerably more time- and labour-intensive than static tests and have been seldom made use of. Flow-through setups to ensure continuous supply of nutrient medium are important for large-scale indoor cultivation of duckweeds (see [47]) but are too complicated and impractical to be of use in routine toxicity testing, especially where downsizing is advantageous.

3.1.3. Data Imaging and Analysis

The simplest means of assessing duckweed growth in toxicity testing is via manual counting and weighing of the fronds. This was specified in the Environment Canada toxicity test (see Table 1) with *Le. minor* (see [31]) and is still sometimes employed to the present day, e.g., [48]. Further examples of manual frond counting have been cited [49], and the weighing of fronds must necessarily be carried out manually. The determination of frond area by means of digital image analysis has a long history: an early example of its use was in the assessment of dose-response relationships between 26 herbicides exhibiting 19 different modes of action on the growth of *Le. paucicostata* [38]. The OECD recommended photographic or digital imaging for recording frond area in its standardized assay procedure [24], which can be conveniently determined with free or commercially available image analysis software. Frond area determined by image analysis for the effects of metal pyrithione antifoulants on *Le. gibba* was found to be up to 10 times more sensitive than visually observed frond number as a toxicity index [50]. However, as indicated by the OECD [24], frond area determination must not necessarily derive from digital imaging: in a recent investigation of the toxicity of boron (B) to *Le. minor*, frond areas were calculated using the image analysis software Image J on photographic images of the fronds [51].

The investigation of 26 different herbicides on the growth of *Le. paucicostata* mentioned above [38] and an early application of the ISO 20079 protocol [25] to test the effect of heavy metals on *Le. minor* [30] used the LemnaTec Scanalyzer and LemnaTec SAW Lemna software (LemnaTec, Würselen, Germany) for the image capture and quantification of total frond area. The use of the LemnaTec Scanalyzer platform to calculate frond number and area in duckweed toxicity testing has been explained in detail [52] and used more recently, e.g., in tests of the effects of the chalcones salsolol A and B and the pseudogualanolide confertin on *Le. paucicostata* [53]. Other examples of image capture and analysis in duckweed toxicity testing are the use of the Nikon Digital Sight DS-5Mc camera with Nikon and Assess image analysis software to study the effect of olive mill wates on *Le. gibba* [39] and of the Canon EOS 450D digital camera and Image J analysis software in investigating the tolerance of *S. polyrhiza* and *Le. minor* to the pesticides atrazine and S-metalochlor [54]. Further software packages for processing digital frond images are the Image Tool analysis program used in the investigation of the toxicity of several antidepressants on *Le. minor* [37], the Lucia 5.0 program used to assess biogenic amine biosynthesis pathway changes associated with dye toxicity to *Le. minor* [55] and the Celleste 6 and Athena software programs marketed by ThermoFisher (Waltham, MA, USA) and IDEA Bio-Medical Ltd. (Rehovot, Israel). An automated microscopy approach to monitoring duckweed growth employing a Zeiss Axiocam 512 color CCD digital camera and Zen Blue 2.6 Professional software to capture fluorescence microscope images of *Le. minor* fronds that were processed by PlantCVv3.10.0 and OpenCVv3.4.9 analysis software may be adaptable for use in duckweed toxicity testing [56].

An interesting example is the use of the Zebrabox (ViewPoint Behavior Technology, Lyon, France), an automated observation and tracking system developed for zebrafish behaviour analysis, to study the growth inhibition potential of rare earth elements on *Wolffia globosa* [49]. Chlorophyll fluorescence detectors have long been used to assess chlorophyll content and chlorophyll fluorescence parameters in duckweed toxicity testing (see Section 4.2.1). The digital images used to determine chlorophyll fluorescence parameters can also be used to calculate the growth parameters of frond number and frond area, as shown in relation to metal and metalloid toxicity to *Le. gibba* [43]. This has the advantage of relating growth phenomena in terms of total frond area to photosynthetically competent leaf matter and the developmental status of the growing fronds [43,57]. A very recent new approach to quantifying duckweed growth in toxicity assessment is the quenching of the light emitted from an LED source by duckweed frond area coverage as registered by a solar cell connected to a digital multimeter. The greater the duckweed frond area produced by growth, the lesser the photoelectric signal emitted from the solar cell [58]. The use of this system was demonstrated with regard to the effects of the herbicides glyphosate and glufosinate and reportedly delivers reliable growth inhibition data within two days.

The optical biosensor system for the measurement of duckweed frond area is reported to be very cost-effective at USD 10 per unit [58]. Other low-cost technical approaches to digital imaging and analysis for objective and quantified biomonitoring have been proposed using standard digital cameras and Aphelion 3.02 framework software. These have been employed using a Nikon Coolpix 995 digital camera in the investigation of the toxicity of non-ionic detergents to *Le. minor* [59] and a Logitech HD Webcam C270 camera for method development with the reference substance potassium dichromate [60]. On a more sophisticated level, a machine learning-based tool (StarDist) simplifies and assists counting processes and helps arrange, manage and calculate inhibition percentages after duckweeds are exposed to contaminants. This was demonstrated in an investigation of the effects of herbicides and rare earths on *Wo. globosa* [49]. Duckbot, a machine customized for automating laboratory research workflows with duckweed, supports the setting up of sample plates with duckweed and growth media and gathering image data, as well as conducting relevant data analysis. Its use was demonstrated in a case study on the effect of salinity on the growth of *Sp. polyrhiza*, *Le. minor* and *Wolffia australiana* [61].

3.2. *Alternatives to the Standardized Tests*

Duckweed toxicity testing can vary fundamentally from the standardized procedures with regard to the test species used as toxicity vehicles and the endpoints determined to define toxicity. The use of these alternatives can also incorporate the downsizing and timesaving modifications of the standardized tests discussed above. Such tests can then be of value for high-volume toxicity testing if they can be shown to compare favourably with the standardized assay procedures.

3.2.1. Alternative Test Organisms

Prior to the establishment of the standardized toxicity tests, a number of duckweed species, including *S. polyrhiza*, *Spirodela oligorhiza*, *La. punctata*, *Le. aequinoctialis*, *Lemna perpusilla* and *Lemna valdiviana*, had been used for investigating the effects of water pollutants on higher plants in addition to *Le. minor* and *Le. gibba* that were singled out as model organisms for the toxicity testing [7]. Some of these duckweeds have still been used occasionally as test organisms in toxicity tests. Employing *Le. aequinoctialis* for testing for water-borne toxicity [62,63] reflects the wish to use a tropical duckweed for use in tropical environments. *S. polyrhiza*, which has a long history as a test organism (see [64]), is used in both classical testing (e.g., [65]) and in newly developed tests using alternative endpoints (see Section 3.2.2). *La. punctata* [66,67] and *Wo. globosa* [49,68] are also used to test for water-borne toxicity on account of their importance in wastewater remediation. Most alternatively used duckweed species have no intrinsic advantages over *Le. minor* and *Le. gibba* as test organisms in traditional assay procedures, but *Wolffia* species would appear to be predestined for use in miniaturized and shortened versions of the standardized assays on account of their very small size and rapid growth.

3.2.2. Alternative Indicators of Toxicity

The standard duckweed toxicity tests determine growth impairment in terms of the number, area, weight or chlorophyll content of whole, mature fronds. However, efforts have also been made to determine detrimental effects on root elongation and juvenile frond expansion that can indicate fundamental disturbances in growth—and hence toxicity—to the test organism. These alterative toxicity endpoints could be of value if they can be assessed as conveniently and accurately as frond growth in the standardized assays. Studies making use of such alternative indicators of toxicity are listed in Table 2.

Two of the alternative indices of toxicity presented in Table 2 constitute root length, either *per se* or upon re-growth after abscission. The three studies that measured the length of intact roots did not reveal any decisive advantage over the established standardized tests. They were limited to testing the effects of heavy metals on *Le. minor* and were based on the experimental methodology of the EC or OECD tests. They thus did not constitute any progress in the downsizing of these test procedures, and in only one instance [70] was a shorter incubation time of 4 days used. The measurement of the intact roots required for the tests is also laborious. However, in two instances [70,71], root length was judged to be a more sensitive and precise indicator of toxicity than frond number that was determined in parallel.

Table 2. Toxicity indicators in tests deviating from the growth parameters used in the standardized toxicity tests listed in Table 1.

Toxicity Indicator	Toxin	Duckweed Species	Reference
Decrease in root length	Ni	*Lemna minor*	[69]
	Pb	*Lemna minor*	[70]
	Cd	*Lemna minor*	[71]
Impaired re-growth of roots after excision	Ag, Cd, Cr, Cu, Hg	*Lemna minor/gibba/paucicostata*	[72]
	atrazine, diuron, paraquat, simazine	*Lemna minor/gibba/paucicostata*	[73]
	Cu	*Lemna minor*	[27]
	Ce	*Lemna minor*	[74]
	B	*Lemna minor*	[51]
	Cu, Cr	*Spirodela polyrhiza*	[75]
	microplastic	*Lemna minor*	[76]
Inhibited growth of first frond produced by germinating turions	22 substances: herbicides, organic/inorganic compounds, metals	*Spirodela polyrhiza*	[64]
	Ni, Cd, Cr	*Spirodela polyrhiza*	[77,78]

The promise of root length as an indicator of toxicity to duckweeds has been much better realized by the measurement of the re-growth of roots from fronds from which the roots were excised prior to the start of the experiment. This was shown in two pioneering studies in which the lengths of roots that re-grew after excision from *Le. minor*, *Le. gibba* and *Le. paucicostata* fronds were measured after exposure to several heavy metals [72] and to several herbicides [73]. These tests constituted drastic miniaturization and shortening of the standardized assay procedures in incubating single rootless fronds in 3 mL medium for 3 days in 24-well plates. They yielded E_c values for root re-growth with the investigated metals and herbicides comparable to those obtained for frond growth. This methodology was later systematically tested for its suitability as a generally applicable test procedure [27]. It was determined to be as sensitive as the conventional ISO 20079 test with the reference toxicant 3,5-dichlorophenol and was shown in national and international interlaboratory comparisons to be reliable and reproducible in tests with $CuSO_4$ and wastewater. This validated *Lemna* root re-growth assay is thus simpler, more rapid, cheaper and more convenient (despite the dexterity required for the frond handling) than the standardized test procedures and is suitable for rapid screening of wastewater and natural waters. The fact that it is carried out with only one de-rooted frond per test for a short time indicates that the frond crowding and nutrient limitation that may be detrimental to the reliability of other miniaturized assays (see above) are probably not at issue here. The procedure was judged to be acceptable as a standardized biological test and for use as a regulatory tool and has accordingly now achieved the status of a standardized duckweed toxicity test in its own right (ISO 4979:2023 [79]). Recent studies have employed the root-re-growth assay as described here to investigate the effects of B on *Le. minor* [51] and of Cu and Cr on *S. polyrhiza* [75]. Despite the advantages of this miniaturized and rapid procedure, some very recent studies have investigated root re-growth upon exposure of *Le. minor* to cerium (Ce) [74] and microplastics [76] using the much more cumbersome methodology of the standardized OECD duckweed toxicity test.

An alternative toxicological indicator related to frond area rather than root length is the growth of the first fronds emerging from germinating turions of *S. polyrhiza*, which are overwintering derivatives of normal growing fronds that spend the winter at the bottom of water bodies in a dormant state and germinate to give rise to normal frond growth in the following spring (see [80]). A microbiotest featuring miniaturization, rapidity and convenience comparable to the root re-growth assays has been developed in which non-dormant *S. polyrhiza* turions are germinated and then incubated individually in 1 mL volumes of medium in 24-well culture plates for 3 days. A decrease in the area of the first frond that protrudes from the geminated turion, which can be digitally recorded and analyzed serves as a measure of toxicity [64]. Due to the investigation of only one germinated turion per test, frond crowding is not an issue and there should not be nutrient depletion sufficient to limit the growth of the first frond protruding from the turion. In addition to being simple and practical to perform, this test is largely independent of the maintenance of duckweed cultures, since the turion starting material can be readily stored until use. The test proved to be of comparable sensitivity to the standardized *Lemna* protocols in tests with 22 substances, including herbicides, organic and inorganic chemicals, and metals. Extensive international interlaboratory comparisons showed the test to be robust and reliable, and it was proposed to and accepted by the ISO as a standard toxicity test (ISO 20227:2017) for duckweeds [64,81]. The test is available as a patented commercial product (Duckweed Toxkit F; MicroBio Tests Inc., Gent, Belgium). It is—like the root re-growth assay—suitable for convenient high-throughput toxicity testing and is particularly useful for ecotoxicology work in the field. Other studies have also investigated the sensitivity of the germination and sprouting of *S. polyrhiza* turions to heavy metals (Ni, Cd and Cr; [77]) and determined that the single frond area produced by germinating turions was the most sensitive test endpoint for the toxic effect of Cd [78]. However, these experiments were carried out in formats resembling the original standardized test protocols and did not recommend systematic test procedures for routine testing.

4. Describing Toxicity: Biomarkers of Toxic Effect

Although the standardized duckweed tests give information as to toxicity itself, it has always been of interest to know what causes the overall detrimental effect to the frond. Toxicity tests themselves have therefore long been accompanied by tests for biomarkers that can help to elucidate the mechanisms underlying the toxicity. Toxic substances that duckweeds can be exposed to include nutrients, metals and organic xenobiotics, the presence of which in natural waters is mostly, but not always, of anthropogenic origin. These substances elicit the generation of "biomarkers of toxic effect" when their influence is correlated with or causally linked to overall observed toxicity [1,82]. Toxicity tests with duckweeds coupled with biomarker-oriented analysis of the test organisms can reveal both the presence and the nature of water contaminant toxicity to aquatic higher plants. However, it should be noted that putatively toxic substance do not always elicit detectable biomarkers of toxic effect! For example, the treatment of *Le. minor* with several concentrations of perfluorooctanoic acid had no discernible effect on biometric frond parameters, chlorophyll content, photosynthetic electron rate or chlorophyll fluorescence parameters [83].

Biomarkers of toxic effect take many forms. Some expressions of toxic influence are readily evident as disturbances to structural and developmental features of duckweeds (see Section 4.1). The "classical" biomarkers of toxic effect include indicators of physiological aberrations of photosynthesis and respiration and are best known as biochemical and metabolic markers related to oxidative stress (Section 4.2). Techniques that are shaping modern biomarker analysis include indications of global alterations in DNA and polymeric structures, but most impressively feature extensive data generation through the application of the various "omics" that can lead to comprehensive understanding of mechanisms of toxicity (Section 4.3).

Biomarkers of toxic effect are in themselves often taken to be proof or verification of toxicity. This is especially evident when toxicity is discussed in terms of such biomarkers without any reference to an actual test of toxicity. An example is the analysis of assumed toxicity of silver (Ag) nanoparticles to *Wo. globosa* in terms of chlorophyll fluorescence parameters (even chlorophyll content is treated as a biomarker), oxidative damage metabolites and antioxidant enzymes [68]. Indeed, a case can be made for the idea that single or multiple biomarkers of effect can be regarded as direct evidence of toxicity, particularly when the effects identified by these biomarkers are obviously seriously detrimental to healthy duckweed growth. Biomarker data compiled during obviously toxic exposure may also correspond well with growth data obtained in the standardized toxicity assays. This was claimed, for example, for chlorophyll fluorescence parameters in comparing the toxicity of four herbicides to three *Lemna* species [73] (although a thorough later investigation of the effect of heavy metals on *Le. gibba* raised doubt about the accuracy of the correspondence [43]). However, biomarkers of effect are often specific for the water contaminant giving rise to them, and thus no particular biomarker can define a toxicity that is due to water contaminants in general, whereas growth inhibition is indeed a quite universal expression of toxicity.

It is preferable to regard the biomarkers being discussed here as being *descriptive* of toxicity, i.e., pointing to what is *going on* in toxicity. In contrast, the growth and chlorophyll synthesis impairment prescribed in the standardized tests are a measure of what has *gone on* in toxicity, i.e., what the result of the "goings-on" actually is. The observation of the biomarkers aims at explaining how the result in terms of growth inhibition came about. However useful, this approach to assessing biomarker data is not necessarily always strictly applicable, as biomarker alterations can take place in the absence of overt toxicity. An example is the detection of metabolite changes shown by ^{1}H-NMR fingerprinting spectra without any macroscopic indication of toxicity in *Le. minor* exposed to several herbicides [84]. In such cases, biomarker changes can indicate impending toxicity at higher water-contaminant dosages.

4.1. Structural and Functional Aberrations

Some anatomical, morphological, developmental and physiological phenomena that may be readily observed in relation to toxicity testing are indicative of functional impairments in duckweeds due to exposure to water contaminants. An early finding was that exposure to heavy metals (Cd, Cu, Ni and Zn) leads to disturbances in the phototactic behaviour of *Le. minor* mesophyll cell chloroplasts [85]. Other aberrant responses of duckweeds under heavy metal stress summarized in [86] include frond chlorosis and developmental disturbances such as frond size reduction, frond abscission and colony disintegration [21,87,88], as well as root rejection and turion formation [89]. Indeed, standardized toxicity tests recommend the routine notation of such features to complement toxicity assessment (e.g., [24]). The floating up and first frond protrusion of germinating turions of *S. polyrhiza* were observed to be delayed in the presence of Cr, Cd and

Ni ions [77,78]. A transition of the chloroplasts of *S. polyrhiza* and *Le. minor* fronds to chloro-amyloplasts and amylo-chloroplasts accompanied by a pronounced accumulation of transitory starch in the presence of Ni and Co was observed by electron microscopy [90,91]. These various phenomena all reflect toxic influences of metals as water contaminants rather than consummated toxicity itself.

4.2. "Classical" Biomarkers

Physiological, biochemical and metabolite biomarkers of toxic effect have long been determined in relation to water contaminant-induced toxicity to duckweeds. A review of biomarkers associated with toxic stress to duckweeds in addition to numerous other, primarily aquatic plants incorporates investigations dating from the end of the last century [82], and a more recent survey presents a comprehensive assessment of biomarkers of toxic effect specifically in duckweeds [1]. An overview of these biomarkers of toxic effect is presented in Table 3. The biomarkers constitute endpoints measured at the end of the incubation periods of toxicity assays. They have been determined in association with the standardized toxicity tests discussed in Section 2; there is as yet no comparable catalogue of biomarkers of toxic effect derived from the more recent standardized tests introduced in Section 3.2.2.

4.2.1. Physiological, Biochemical and Metabolite Endpoint Biomarkers

An in-depth analysis of toxicity-related biomarkers is beyond the scope of this review (see [1,82] in this regard), but two complexes of biomarkers that are frequently documented are of particular interest. Many of the biomarkers listed in Table 3 are related to the wide-reaching toxic effects of oxidative stress that are frequently occasioned by exposure of duckweeds to water contaminants. This stress results from the production of reactive oxygen species (ROS), which represent the most important chemical biomarkers of toxic effect and lead to the production of metabolites such as the lipid peroxidation product MDA. In response, antioxidative solutes, such as glutathione and ascorbate, and antioxidative enzymes, such as peroxidase and catalase, are formed to counteract the effects of the ROS; Phase I and II enzymes are induced to detoxify water contaminants leading to oxidative stress (see [1,81] for detailed discussion). An interesting parallel to this elicitor–response scenario is the finding that biogenic amines—in particular, tyramine and spermine—accumulated significantly in *Le. minor* in response to toxic exposure to the antibiotic tetracycline [92]. A subsequent investigation illustrated the response of the biogenic amine biosynthesis pathway of the duckweed to toxic exposure to the dye gentian violet in terms of a stimulation of tyrosine decarboxylase (leading to tyramine accumulation) and inhibition of s-adenosylmethionine and ornithine decarboxylases [55]. Since the biomarkers under discussion here have been determined and compiled since the establishment of the standardized toxicity tests, they are not new in the sense of this review but rather represent the *status quo* of classical toxicity test results. They continue to be routinely determined up to the present: studies on the effects of norfloxacin on *S. polyrhiza* [65] and benzalkonium chloride on *Le. minor* [93] are just two recent examples.

Table 3. Examples of "classical" biomarkers determined to accompany toxicity to duckweeds exposed to various metals and organic xenobiotics. The "examples" and "duckweed species tested" are assigned to the respective water contaminants; the right-hand column summarily lists the biomarkers determined in such investigations. Content taken from [1,82], in which the data summarized here are comprehensively presented and discussed. MDA: malondialdehyde, SOD: superoxide dismutase.

<table>
<tr><th>Water Contaminant</th><th>Examples</th><th>Duckweed Species Tested</th><th>Biomarkers</th></tr>
<tr><td colspan="3">Metals</td><td rowspan="12">• altered photosynthesis pigment contents/ratios
• perturbations of chloroplast ultrastrucure
• altered chlorophyll fluorescence parameters
• mpaired photosynthesis photochemistry
• inhibited photosynthetic O_2-evolution
• increased reactive oxygen species (ROS) content
• increase in lipid peroxidation products (MDA)
• increased Phase I enzyme (functionalizing: e.g., peroxidases, esterase) activities
• increased Phase II enzyme (conjugating: e.g. glutathione-S-transferase, methyl-transferase) activities
• increased antioxidant enzyme (e.g. peroxidases, catalase, SOD) activities
• increased antioxidant solute (e.g. ascorbate, glutathione, thiols) concentrations
• increased contents of phenols, sterols and flavonoids
• increases in starch synthesis enzyme activities
• starch accumulation</td></tr>
<tr><td>Heavy metals</td><td>Cd, Co, Cu, Cr, Hg, Zn, Ni, Pb</td><td>S. polyrhiza
La. punctata
Le. minor</td></tr>
<tr><td>Nanoparticles</td><td>Ag, Cu</td><td>S. polyrhiza
Le. minor
Wo. globsa</td></tr>
<tr><td>Metalloids</td><td>As, B, Se</td><td>La. punctata
Le. minor/gibba
Wo. arrhiza</td></tr>
<tr><td>Rare earths</td><td>Pr, Ce</td><td>S. polyrhiza</td></tr>
<tr><td colspan="3">Organic xenobiotics</td></tr>
<tr><td>Natural chemicals:
bacterial toxins
coal tar hydrocarbons</td><td>
coronatine
fluoranthine</td><td>
Le. paucicostata
Le. minor</td></tr>
<tr><td>Industrial chemicals:
solvents
dyes
surfactants</td><td>
diethyl phthalate
Malachite Green
alkyl dimethyl amine oxide</td><td>
S. polyrhiza
Le. minor
Le. minor</td></tr>
<tr><td>Agricultural chemicals:
herbicides
insectucides
fungicides
growth retardants
anti-ozonants</td><td>
atrazine
teflubenzuron
epoxiconazole
uniconazole
ethylene diurea</td><td>
Le. minor
Le. minor
Le. minor
La. punctata
Le. minor</td></tr>
<tr><td>Pharmaceuticals:
analgesics
antibiotics
anti-depressants</td><td>
diclofenac
tetracycline
fluoxetine</td><td>
Le. minor
Le. minor
Le. minor</td></tr>
<tr><td>Particles:</td><td>polyethylene microbeads</td><td>Le. minor</td></tr>
</table>

Chlorophyll fluorescence also plays an important role as a "classical" biomarker of toxic effect in duckweeds. It is light re-emitted by light-excited chlorophyll molecules that return to non-excited states and is—as a non-invasive measurement of photosystem II (PSII) activity—an indicator of photosynthetic energy conversion in plants. Chlorophyll fluorometer protocols measure the photosynthetic efficiency of PSII, both in the light (ΔF/Fm′) and in a dark-adapted state (Fv/Fm) in terms of multiple fluorescence parameters (see [94]). Chorophyll fluorescence-based bioassays for rapid, high-throughput screening of toxic effects on the photosynthesis of aquatic macrophytes were introduced soon after the establishment of the standardized duckweed toxicity tests [95,96]. They have played an important role in duckweed toxicity testing ever since (see Table 3) and remain at the forefront of the characterization of toxic effects of water contaminants on duckweeds.

Chlorophyll fluorescence imaging-based phenotyping of *S. polyrhiza* exposed to Ni, Cr and NaCl showed that nine calculated fluorescence parameters exhibited very divergent sensitivities to dark- and light-adapted states of photosystem II activity, indicating the importance of assessing several chlorophyll fluorescence parameters in relation to toxicity [94]. This study also indicated that changes in duckweed photochemical efficiency do not necessarily correspond to growth responses. On the other hand, the exposure of *Le. minor* and *S. polyrhiza* to 3, 30 and 600 mg dimethyl phthalate (DMP)/L showed that multiple chlorophyll fluorescence parameters exhibited significant alteration only at the highest DMP concentration, in accordance with reduced chlorophyll accumulation as a standard toxicity index at this concentration only [97]. This study also highlighted the utility of leaf reflectance spectra, which reflect both leaf surface and internal structure and pigment characteristics, as biomarkers complementary to chlorophyll fluorescence parameters. The values of six spectral reflectance indices only varied significantly in concord with fluorescence parameter changes at the highest DPM concentration. In addition, the chlorophyll fluorescence imaging technique allowed the assessment of spatial variations in the distribution of chlorophyll fluorescence within the leaves. This has been further expanded upon by the use of chlorophyll fluorescence imaging to demonstrate the functional heterogeneity of fronds and colonies of *S. polyrhiza*, *La. punctata* and *Le. minor* treated with Ni and Cr [57].

4.2.2. The Progression of Toxic Effects in Time

Most biomarkers of toxic effect are determined at defined times after the onset of treatment: at 1 to 7 days, as discussed above. Monitoring the appearance or development of such biomarkers within these timespans would permit judgement as to the progression of toxicity along a timescale. This can be important to judge whether standardized toxicological endpoint measurements reflect continual ongoing toxicity or if they result from toxic processes that are either belatedly or no longer effective.

Any non-invasive measurement regime leaving the test material intact, i.e., digital frond imaging with CCD cameras or chlorophyll fluorescence detectors, is suitable for repeated timeline measurements. An early example of this is the digital capture of *Le. paucicostata* frond images at daily intervals during the week-long exposure of the duckweed to numerous herbicides [38]. Visualization of the short-term progression of oxidative stress in *Le. minor* was described in 2007, whereby emissions of a fluorescent probe due to reactive oxygen species production elicited by the vitamin menadione were captured by an electron-multiplying CCD camera at intervals over a period of 6 h [98]. Although this was claimed to represent the observation of real-time toxicity, the continual CO_2 uptake/emission measurements made with *Le. minor* exposed to four herbicides and four metals using a respiratory activity measuring system based on infrared (IR) technology over a 24 h period [99] more accurately reflect the progress of toxicity in real time. CO_2 exchange rates were stimulated at low concentrations of the test substances and mostly inhibited at high concentrations. Short-term repetitive or continuous measurement of oxidative stress or CO_2 exchange rates could accordingly be promising for initial screening for potential toxicity. Interestingly, the IR spectroscopy that was used to quantify decarboxylation reactions in biogenic amine biosynthesis [55] could also be used to monitor numerous toxicity-related enzyme reactions by determining absorbance changes in reactants over time [100,101].

Chlorophyll fluorescence parameters are most promising for repeated analysis in real time using the commercially available Imaging-PAM and FluorCam platforms (see [94]). Indeed, an Ecotox Photosystem Tool has recently been developed that combines the use of an Imaging PAM CCD camera and an IR gas analyser to enable simultaneous and continuous measurements and analysis of leaf area, CO_2 exchange and chlorophyll fluorescence parameters [102]; see also [97]. Such a sophisticated system enables non-destructive analysis (allowing further investigation of the test material) that can detect multiple early toxicity symptoms even if no macroscopic indications of toxicity are evident.

4.3. Modern Methods of Biomarker Analysis

In recent years, sophisticated techniques for analysing transcripts, proteins and metabolites from plant material have made possible the determination of large numbers of biomarkers of toxic effect in duckweed tissues. These techniques can reveal numerous toxicity-associated biomarkers that are complementary to individual "classical" biomarkers that may have been detected. Such biomarkers constitute the "omics", or the collective characterization and quantification of entire sets of biological molecules that translate into the structure, function and dynamics of an organism. The application of transcriptomics, proteomics, metabolomics and lipidomics to duckweed toxicity testing has greatly expanded the possibilities of achieving a comprehensive understanding of mechanisms underlying toxicity. This development has been complemented by the possibility of assessing changes in complex polymeric structures—including cellular DNA—that are associated with toxicity to duckweeds. Transformation has also become a tool in toxicity research in duckweeds, especially with respect to phytoremediation. Table 4 gives an overview of the investigations that have been carried out in these regards. The integration of the biomarker data available from these various modern approaches can throw comprehensive new light on toxicity to duckweeds.

4.3.1. Transcriptomics

The microarrays that were in early use to detect differential gene expression in plant material were not made use of in duckweed toxicity testing, but modern transcriptomic analysis is now being applied in this regard. High-throughput or next-generation sequencing enables DNA and RNA to be sequenced in a rapid and cost-effective manner. It can reveal the differential gene expression that takes place in duckweeds in response to stimuli or triggers and leads to the overall response of the organism. It has been useful in this regard for understanding how duckweeds react and respond to stresses elicited by, e.g., ionizing radiation in *Le. minor* [116], salinity in *S. polyrhiza* [117], heat in *S. polyhiza* [118] and N starvation in *Le. aequinoctialis* [119]. It has also been used to analyse phytohormone-triggered developmental changes in duckweeds, e.g., salicylic acid-induced flowering in *Le. gibba* [120] and abscisic acid-induced turion formation in *S. polyrhiza* [121] (see also [122]). Transcriptomics has, however, only seldom been applied to the determination of toxicity due to water contaminants in duckweeds.

Table 4. Investigations carried out to describe toxicity to duckweeds in contemporary approaches complementary to the acquisition of the "classical" biomarkers of effect referred to in Section 4.2. The results of the individual studies are discussed in the text.

Type of Investigation	Toxin	Duckweed Species	Reference
Transcriptomics	NH_4^+	*Lemna minor*	[103]
	Cd	*Landoltia punctata*	[104]
	heavy metal mixture	*Spirodela polyrhiza*	[105]
	atorvastatin/bentazon	*Lemna minor*	[106]
	imazamox	*Lemna minor*	[29]
Proteomics	Al	*Lemna minor*	[107]
	atorvastatin/bentazon	*Lemna minor*	[106]
Metabolomics	herbicides, phytotoxin	*Lemna minor*	[84]
	herbicides	*Lemna minor*	[108]
	diclofenac	*Lemna minor*	[109]
Lipidomics	heavy metal mixture	*Spirodela polyrhiza*	[105]
FTIR spectroscpy	herbicides, heavy metals	*Lemna minor*	[110]
	fluoralkyls	*Lemna minor*	[111]
Genotoxicity	Hg	*Lemna minor*	[112]
	Zn-MP/NP	*Lemna minor*	[113]
	dimethyl phthalate	*Spirodela polyrhiza*	[114]
Transformation	Cd	*Lemna turionifera*	[115]
	heavy metal mixture	*Spirodela polyrhiza*	[105]

The inhibitory action of high concentrations of ammonium on the growth of *Le. minor* [123] was found to correspond to over 14,000 differentially expressed genes (DEGs) [103]. Most of the DEGs were downregulated under NH_4^+ toxicity. However, genes required for lignin biosynthesis in the phenylpropanoid biosynthesis pathway were upregulated, which indicates a strengthening of the cell wall. Genes related to programmed cell death were also upregulated, as were genes involved in the production of both reactive oxygen species (ROS) and the antioxidant enzyme system to scavenge the ROS.

The heavy metal cadmium (Cd) is a potent inhibitor of the growth of duckweeds such as *Le. minor* and *S. polyrhiza* [30,78]. A transcriptomic analysis of *La. punctata* to investigate the relative tolerance of this species to Cd [104] detected over 9000 DEGs associated with Cd toxicity, including genes associated with DNA repair and carbohydrate metabolic flux. Transcripts encoding enzymes involved in cell wall biosynthesis were downregulated, while the transcription levels of genes involved in glycolysis and the tricarboxylic acid cycle were upregulated. Upregulated transcripts encoded enzymes involved in sulphur assimilation and in antioxidation to combat ROS metabolism and led to enhanced expression of tonoplast-localized transporters to facilitate vacuolar accumulation of Cd. In a recent study of the tolerance of *S. polyrhiza* to heavy metal stress [105] (see also Section 4.5), the expression of genes involved in fatty acid biosynthesis, β-oxidation and lipid degradation was shown to be upregulated to enable the incorporation of oxidative lipid degradation products into triacylglycerides to prevent cellular lipotoxicity due to the presence of unsaturated free fatty acids. A particularly highly expressed transcription factor was also indicated to be important in imparting heavy metal tolerance (see Section 4.5).

Transcriptomics have also been employed in two recent studies examining the toxic effects of organic xenobiotics on duckweeds. In an examination of the influence of the pharmaceutical atorvastatin and the herbicide bentazon on *Le. minor* [106] (see also Section 4.3.2), the herbicide resulted in a far greater number of DEGs than did the pharmaceutical at low effective growth-inhibitory concentrations. Highly regulated genes of the ethylene-activated signalling pathway and of lipid and glycerophospholipid metabolism were differentially expressed only in response to atorvastatin, while the expression of highly regulated genes of cellular responses to light stimuli and to photosynthesis was affected only by exposure to bentazon, thus giving an indication of how specific the underlying mechanisms of toxicity can be with the elicitor. In an investigation of the mechanism of enantioselective toxicity, racemic- (Rac-), R- and S-preparations of the chiral imidazolinone pesticide imazamox (IM) were applied to *Le. minor* [29]. Transcriptional analysis showed that the treatments with Rac-, R- and S-IM produced 3107, 2580 and 13 DEGs, respectively, which were consistent with the comparative extent of toxicity of the three preparations to the duckweed. The DEGs resulting from Rac- and R-IM administration were related mainly to photosynthetic carbon assimilation and antenna proteins, glutathione metabolism, the pentose phosphate pathway, hormone signal transduction and zeatin biosynthesis. The DEGs in the samples treated with S-IM were specifically related to phenylalanine metabolism and phenylpropanoid biosynthesis. This study illustrates the specific toxicity and effects on transcription that enantiomers can exert.

Transcriptomic analyses of the effects of water contaminants of duckweeds are major projects that are not realistic options for routine and high-sample-volume toxicity testing. Their value for elucidating the toxicity mechanisms related to selected water contaminants, as illustrated above is, however, evident, and they should accompany the toxicity testing of as many substances as is practical for better understanding of both elicitor- and duckweed-specific toxicity. The comprehensive understanding of the gene expression involved in water-borne toxicity to duckweeds afforded by transcriptomic studies can be complemented by the use of further "omics" to illustrate the proteinaceous and metabolic consequences of the differential gene expression having taken place.

4.3.2. Proteomics

Proteomics is a powerful tool for understanding changes in the plant proteome under toxic stress. Changes in the protein composition of duckweed fronds induced by exposure to water contaminants show the result of differential gene expression due to toxic influences. This can be determined by the identification of tryptic peptides of proteins extracted from control and stressed duckweed fronds by mass spectrometry and the quantitative relation of their spectra to known protein structures. Using this approach, an analysis of the effect of 15 days of exposure of *Le. minor* to aluminium (Al) detected 261 differentially expressed proteins (DEPs) identified by MS/MS parallel to growth inhibition and chlorosis of the fronds [107]. Most of the DEGs related to the citrate cycle and amino acid metabolism were upregulated, whereas those associated with energy metabolism and glyoxylate and dicarboxylate metabolism were predominantly downregulated. Antioxidant enzyme-related proteins were important in the response to Al.

A recent transcriptomic analysis of the toxic action of the pharmaceutical atorvastatin and the herbicide bentazon on *Le. minor* showed up to 1200 (80% downregulated) and 2700 DEGs (60% downregulated), respectively, depending on the elicitor concentrations employed [106] (see also Section 4.3.1). A complementary proteomic investigation showed that the regulation directions of the DEGs corresponded well to changes in the concentrations of the respective actual proteins, showing that DEGs can truly reflect changes in protein composition (and *vice versa*). An interesting aspect of this study was that toxicity was determined preliminarily over a period of 7 days with a standardized test, whereupon the sampling of the fronds for the transcriptomic and proteomic investigations was carried out after only three days of exposure to the toxins and thus during the development of the eventually determined toxicity.

4.3.3. Metabolomics

Metabolomics is important for understanding the composition and changes in the plant metabolome under toxic stress. The metabolome of a plant at any one time represents the entirety of the small-molecule substrates, intermediates and products of cell metabolism present in the tissues, providing a direct functional readout of the physiological state of the plant material. Metabolomics can thus relate global changes in metabolite composition and changes in particular groups of metabolites to the detrimental impacts of water-borne substances on duckweeds revealed by toxicity tests. In an early attempt at this, ^{1}H NMR fingerprints of extracts of *Le. minor* exposed to several different herbicides and a phytotoxin revealed elicitor-specific differences in metabolic composition due mainly due to differences in substances containing methylene, methine, hydroxy, amine, thiol, olefin and aldimine groups [84].

More recently, comprehensive metabolomic analyses of toxicity in duckweeds have been facilitated by the characterization of plant extracts by gas or liquid chromatography coupled with mass spectrometry. GC/EI/MS has been used to show that the application of the herbicides metribuzin and glyphosate and their mixtures to *Le. minor* results in an increase in the amino acid pool, thus indicating increased proteolytic activity, and in fluctuations in metabolites such as salicylate, caffeate, γ-aminobutyric acid, trehalose and squalene that indicate the activation of pathways involved in the relief of stress [108]. The anti-inflammatory drug diclofenac was shown by means of untargeted screening with RPLC-HILIC-ESI-TOF-MS to result in changes in organic acids, lignin, sugars, amino acids, dipeptides, flavonoids and fatty acids in *Lemna minor* that can be related to the activation of various defence mechanisms to ameliorate stress conditions. A targeted analysis of amino acid changes illustrated that different drug concentrations differentially influenced amino acid synthesis pathways [109].

The metabolome is quantitatively related to the other cellular ensembles of the genome, transcriptome, proteome and lipidome. The integration of metabolomics with other omics information can provide a better understanding of toxicity to duckweeds in the sense of systems biology. One recent attempt at this approach is the integration of transcriptomics and lipidomics (lipid-related metabolomics) in the investigation of the toxic effects of wastewater heavy metals on *S. polyrhiza* [105] (see also Section 4.3.1). This study revealed that the heavy metal treatment stimulated degradation of the mono- and digalactosyldiacylglyerol membrane lipids, resulting in the accumulation of toxic unsaturated free fatty acids, and that these harmful degradation products were removed by incorporation (and thus sequestration) into triacylglycerol. Parallel to these lipidomic findings, transcriptomic analysis revealed that genes involved in fatty acid synthesis, β-oxidation and lipid degradation were upregulated, while genes involved in cuticular wax biosynthesis were downregulated by the heavy metal treatment. The combination of metabolomics and transcriptomics led to a comprehensive understanding of the effects of heavy metals on the lipid metabolism of *Le. minor*. In the same vein, the complementation of transcriptomics by proteomics was important for the understanding of the influence of atorvastatin and bentazon on *Le. minor* referred to in Sections 4.3.1 and 4.3.2 [106].

4.3.4. FTIR Spectroscopy

Fourier-Transform Infrared Spectroscopy (FTIRS) of duckweed fronds exposed to toxic substances has enabled the detection of changes in complex polymeric molecules as well as in metabolites associated with toxicity. FTIRS revealed biochemical alterations attributable to structural protein, lipid, nucleic acid and carbohydrate changes associated with the toxicity of Cu, Cr, the herbicides atrazine and acetochlor, and heavy metal-containing industrial wastewaters to *Le. minor* [110]. In a similar vein, FTIRS showed that significant structural and functional alterations were induced in *Le. minor* at the biochemical level upon exposure to four per- and polyfluoroalkyl substances (PFASs) [111]. Even at PFAS concentrations that did not visibly inhibit growth, lipid peroxidation, protein aggregation, and protein and DNA conformation changes were observed: alterations in lipid, protein and DNA were concentration-related and compound-specific. This would make FTIRS a most sensitive indicator of even potential water-borne toxicity to duckweeds.

4.4. Genotoxicity

Are the toxic effects of water pollutants due merely to altered transcription and the consequential changes in enzyme activities and metabolite levels, or is damage to the DNA itself involved? While numerous environmental pollutants indeed cause damage to plant DNA [124], research on the impact of water contaminants on aquatic organisms has focused primarily on the physiological effects of the pollutants, while their genetic damage potential remains poorly understood [125]. The investigation of water-contaminant-related genotoxicity in duckweeds is still in its infancy.

DNA damage was indicated to accompany the effects of reduced photosynthetic pigment synthesis and increases in oxidative stress markers and antioxidant enzyme activities and metabolite concentrations in *Le. minor* exposed to mercury (Hg). The RAPD (random amplified polymorphic DNA) profile of the treated plants exhibited new bands and missing normal DNA amplicons, indicating a decline in genomic template stability [112]. The Comet assay that has been employed as a standard technique for evaluation of DNA damage/repair, biomonitoring and genotoxicity testing in animal research since the 1980s has only recently been used to detect DNA damage in both *Le. minor* and *S. polyrhiza* upon treatment with dimethyl phthalate (DMP) [114] and in *Le. minor* upon exposure to micro- and nano-zinc (Zn) particles [113]. The detection of the genotoxic effect of DMP as soon as 24 h after the onset of treatment indicates that DNA damage can provide a very rapid indication of severe toxic effects. The DNA damage was considered in both studies to result from the impact of reactive oxygen species (ROS) formed as a response to the DMP and Zn particle treatments, in agreement with the view that DNA damage quite generally results from oxidative stress [124]. The routine investigation of potential genotoxicity could be profitably integrated into toxicity testing with duckweeds to provide a comprehensive overview of the damage to DNA that water contaminants can exert on duckweeds as representatives of higher aquatic plants. This could be most practically realized by extensive use of the widely used Comet assay that is a versatile, sensitive and relatively straightforward procedure that allows for high-throughput screening [125]. Since the Comet assay indicates reversible DNA damage [125], it provides the opportunity to better understand metabolic responses to water-borne toxicants in terms of how these responses (e.g., the enhancement of antioxidative solute concentrations and enzyme activities) may relate to the alleviation of conditions leading to DNA damage (i.e., the presence of high reactive oxygen species contents) to allow DNA repair to take place.

4.5. Transformation

The diagnostic value of biomarkers identified in relation to toxicities to duckweeds may be enhanced in some cases by integrating transformation procedures into the analysis of toxicity. This has recently been the case in two examinations of heavy metal toxicity primarily concerned with understanding and improving the tolerance of duckweeds to heavy metals. Tolerance to heavy metals—or other water contaminants—is an important property of duckweeds in their use in the remediation of contaminated wastewater or natural waters, which is an important opportunity for putting duckweeds to good use [16,17,126]. All manner of water contaminants can be removed from water by duckweeds, the versatility of *Le. minor* in this regard being particularly impressive [127]. A typical recent environmentally relevant example illustrates the potential of a duckweed (*Le. aequinoctialis*) to remove contaminants (the metals Cd, Cr, Pb and V) from artificial wetland water [128]. The remediative ability of duckweeds presupposes the ability of the macrophytes to tolerate certain levels of the toxicants that are to be taken up: increased tolerance enhances the remediative potential.

In one case, the biomarker providing an impetus for the transformation was a metabolite, in the other a gene expression product. It is of note that neither of these studies employed a toxicity test per se to document the heavy metal toxicity underlying tolerance phenomena (although some indications of toxicity compatible with such tests were provided): they illustrate the use of biomarkers to define toxicity (see Section **4**).

One of the metabolites found to respond to Cd stress in *Lemna turionifera* (indicated morphologically by increased root abscission) is glutamate (Glu), the content of which increased and activated the Ca^{2+} signalling pathway in fronds [129]. Phosphoserine aminotransferase (*PSAT*), a crucial enzyme in Glu metabolism, from *Arabidopsis thaliana* was introduced into *Le. turionifera*, and the response of the transformed duckweed to exposure to Cd was examined [115]. The *PSAT*-transgenic duckweed proved to be better-suited to tolerate Cd stress than the wild type, as indicated by prevention of root abscission, increased photosynthetic capacity and pigment content, and upregulated antioxidant enzyme activities. In addition, an upregulation of Glu metabolism-related enzymes and an increased Glu content were determined by transcriptomic and metabolomic analyses. The observation of Glu as a Cd toxicity-related biomarker thus led to the insight of the probable importance of Glu metabolism for heavy metal tolerance in aquatic plants.

A transformational offshoot of the transcriptomic identification of a biomarker was a valuable component of a study providing insight into the mechanisms of heavy metal stress in *S. polyrhiza* [105]. Treatment of the duckweed with heavy metals in flue gas desulphurization (FGD) wastewater resulted in reduced growth, chlorosis and decreased chlorophyll content of the fronds, which are indeed criteria of toxicity according to the standardized duckweed toxicity assays. Lipidomic and transcriptomic investigations revealed reduced membrane lipid biosynthesis and stimulated lipid degradation resulting in triacylglycerol and carbohydrate accumulation. Fatty acid biosynthesis, β-oxidation and lipid degradation-related genes were upregulated, in contrast to downregulated cuticular wax biosynthesis genes. Of particular interest was the identification of a highly upregulated transcription factor gene (SpWRI3), the ecotopic expression of which in *A. thaliana* led to increased tolerance of the transgenic plant to FGD wastewater (lower malondialdehyde and higher glutathione levels). This indicates that SpWRI3 is important for the tolerance of *S. polyrhiza* to heavy metals.

The insights into water contaminant tolerance gained from these studies suggest that targeted transformation of duckweeds, for which protocols have been developed (see [8,126]), can be an effective tool to improve the water remediation capacity of the macrophytes. An example is the overexpression of the photorespiratory pathway gene serine: glyoxylate aminotransferase from *A. thaliana* that led to salt stress tolerance in *Le. minor* [130]. Transformation can be a useful tool for understanding toxic scenarios, as illustrated in the two examples detailed above, but the advisability of its actual implementation can only be decided upon on the basis of appropriate toxicity indices on an individual basis. It is not suitable for routine introduction into duckweed toxicity testing, as it does not reveal the results of toxicity itself but is rather a specifically defined genetic manipulation useful for the interpretation, and possibly the exploitation, of these results. Transformation is not the only molecular tool that can be used to this end. The genetic constitution of duckweeds may also be able to be altered to improve the water-remediative attributes of the macrophytes by techniques such as the artificial miRNA gene silencing shown to suppress the expression of the magnesium chelatase subunit *CH42* that resulted in a reduction in chlorophyll pigmentation in *Le. minor* [131]. Another promising technique is the CRISPR/Cas9-mediated gene editing [132] that produced biallelic mutations with an albino phenotype in *Le. aequinoctialis* [133] in its only application to date in duckweeds.

5. Conclusions

The standardized tests that were established between 1991 and 2006 remain popular to the present and are still quite adequate for determining water-borne toxicity to duckweeds as representatives of higher aquatic plant life. Experience has shown that duckweed species other than *Le. minor* and *Le. gibba*—especially *S. polyrhiza*—are also suitable for toxicity testing. Efforts to downsize and shorten the test procedures have been most successful with the development of new standardized toxicity tests using the endpoints of root re-growth and first-frond growth immediately following turion germination. The classical metabolic, enzymatic and physiological biomarkers that have long been determined in the context of toxicity to duckweeds are still valuable descriptors of water-borne toxicity, and progress has been made in monitoring them along time-lines. However, recent developments in transcriptomics, proteomics and metabolomics—complemented by genotoxicity assays and transformation approaches—can provide much more toxicity-related data that are invaluable for the understanding of the mechanisms underlying toxicity. In an earlier paper, a proposal was made for the identification of unknown toxicants in contaminated water samples on the basis of toxicant-biomarker databases [1]. The realization of this goal becomes ever more feasible with the wealth of biomarkers of toxic effect that can be made available by the modern approaches to duckweed toxicity description described in this review.

Funding: This research received no external funding.

Institutional Review Board Statement: Not applicable.

Informed Consent Statement: Not applicable.

Data Availability Statement: No new data were created or analyzed in this study.

Conflicts of Interest: The author declares no conflicts of interest.

References

1. Ziegler, P.; Sree, K.S.; Appenroth, K.-J. Duckweed biomarkers for identifying toxic water contaminants? *Environ. Sci. Pollut. Res.* **2018**, *26*, 14797–14822. [CrossRef] [PubMed]
2. Ceschin, S.; Bellini, A.; Scalici, M. Aquatic plants and ecotoxicological assessment in freshwater ecosystems: A review. *Environ. Sci. Pollut. Res.* **2021**, *28*, 4975–4988. [CrossRef]
3. Lewis, M.A. Use of freshwater plants for phytotoxicity testing: A review. *Environ. Pollut.* **1995**, *87*, 319–326. [CrossRef]
4. Sculthorpe, C.D. *The Biology of Aquatic Vascular Plants*; Edward Arnold: London, UK, 1967; 610p.

5. Landolt, E. *The Family of Lemnaceae—A Monographic Study, Vol. 1. Biosystematic Investigations in the Family of Duckweeds (Lemnaceae) (Vol. 2)*; 71. Heft; Veröffentlichungen des Geobotanischen Institutes der Eidgenössischen Technischen Hochschule, Stiftung Rübel: Zürich, Switzerland, 1986.
6. Bog, M.; Appenroth, K.-J.; Sree, K.S. Key to the determination of taxa of Lemnaceae: An update. *Nord. J. Bot.* **2020**, *38*, e02658. [CrossRef]
7. Wang, W. Literature review on duckweed toxicity testing. *Environ. Res.* **1990**, *52*, 7–22. [CrossRef] [PubMed]
8. Acosta, K.; Appenroth, K.J.; Borisjuk, L.; Edelman, M.; Heinig, U.; Jansen, M.A.K.; Oyama, T.; Pasaribu, B.; Schubert, I.; Sorrels, S.; et al. Return of the Lemnaceae: Duckweed as a model plant system in the genomics and postgenomics era. *Plant Cell* **2021**, *33*, 3207–3234. [CrossRef]
9. Bog, M.; Appenroth, K.-J.; Sree, K.S. Duckweed (Lemnaceae): Its molecular taxonomy. *Front. Sustain. Food Syst.* **2019**, *3*, 117. [CrossRef]
10. Bog, M.; Sree, K.S.; Fuchs, J.; Hoang, P.T.N.; Schubert, I.; Kuever, J.; Rabenstein, A.; Paolacci, S.; Jansen, M.A.K.; Appenroth, K.-J. A taxonomic revision of *Lemna* sect. *Uninerves (Lemnaceae). Taxon* **2020**, *69*, 56–66. [CrossRef]
11. Braglia, L.; Lauria, M.; Appenroth, K.-J.; Bog, M.; Breviario, D.; Grasso, A.; Gavazzi, F.; Morell, L. Duckweed species genotyping and interspecific hybrid discovery by tubulin-based polymorphism fingerprinting. *Front. Plant Sci.* **2021**, *12*, 625670. [CrossRef]
12. Braglia, L.; Ceschin, S.; Iannelli, M.A.; Bog, M.; Fabriani, M.; Frugis, G.; Gavazzi, F.; Giani, S.; Mariani, F.; Muzzi, M.; et al. Characterization of the cryptic interspecific hybrid *Lemna* × *mediterranea* by an integrated approach provides new insights into duckweed diversity. *J. Exp. Bot.* **2024**, *75*, 3092–3110. [CrossRef]
13. Appenroth, K.-J.; Jansen, M.A.K.; Lam, E.; Shoham, T.; Sree, K.S. Important terms in duckweed research defined. *Duckweed Forum (ISCDRA Newsl.)* **2024**, *12*, 90–94. Available online: http://www.ruduckweed.org/ (accessed on 14 October 2024).
14. Romano, L.E.; Braglia, L.; Iannelli, M.A.; Lee, Y.; Morello, L. A survey of duckweed species in southern Italy provided first occurrences of the hybrid *Lemna* × *mediterranea* in the wild. *bioRxiv preprint* **2024**. [CrossRef]
15. Tippery, N.P.; Les, D.H. Tiny plants with enormous potential: Phylogeny and evolution of duckweeds. In *The Duckweed Genomes*; Cao, X., Fourounjian, P., Wang, W., Eds.; Compendium of Plant Genomes; Springer: Cham, Switzerland, 2020. [CrossRef]
16. Ziegler, P.; Sree, K.S.; Appenroth, K.-J. Duckweeds for water remediation and toxicity testing. *Toxicol. Environ. Chem.* **2016**, *98*, 1127–1154. [CrossRef]
17. Ziegler, P.; Sree, K.S.; Appenroth, K.-J. The uses of duckweed in relation to water remediation. *Desalin. Water Treat.* **2017**, *63*, 327–342. [CrossRef]
18. Ziegler, P.; Adelmann, K.; Zimmer, S.; Schmidt, C.; Appenroth, K.-J. Relative in vitro growth rates of duckweeds (Lemnaceae) – The most rapidly growing higher plants. *Plant Biol.* **2015**, *17* (Suppl. 1), 33–41. [CrossRef]
19. Sree, K.S.; Sudakaran, S.; Appenroth, K.-J. How fast can angiosperms grow? Species and clonal diversity of growth rates in the genus *Wollfia* (Lemnaceae). *Acta Physiol. Plant.* **2015**, *37*, 204. [CrossRef]
20. Kandeler, R. Überlebensstrategien bei Wasserlinsen. *Biol. Rundsch.* **1988**, *26*, 347–354.
21. Topp, C.; Henke, R.; Keresztes, A.; Fischer, W.; Eberius, M.; Appenroth, K.J. A novel mechanism of abscission in fronds of *Lemna minor* L. and the effect of silver ions. *Plant Biol.* **2011**, *13*, 517–523. [CrossRef]
22. Kim, I. Structural differentiation of the connective stalk in *Spirodela polyrhiza* (L.) Schleiden. *Appl. Microsc.* **2016**, *46*, 83–88. [CrossRef]
23. Laird, R.A.; Barks, P.M. Skimming the surface: Duckweed as a model system in ecology and evolution. *Am. J. Bot.* **2018**, *105*, 1962–1966. [CrossRef]
24. OECD (Organisation for Economic Co-operation and Development). Test No. 221: *Lemna* sp. Growth Inhibition Test. In *OECD Guidelines for the Testing of Chemicals*; Section 2; OECD Publishing: Paris, France, 2006. [CrossRef]
25. *ISO 20079:2005*; Water Quality—Determination of the Toxic Effect of Water Constituents and Wastewater on Duckweed (Lemna minor)—Duckweed Growth Inhibition Test. International Organization for Standardization: Geneva, Switzerland, 2005. Available online: https://www.iso.org/standard/34074.html (accessed on 15 January 2025).
26. USEPA (United States Environmental Protection Agency). *Aquatic Plant Toxicity Test Using Lemna spp. Ecological Effects Test Guidelines OPPTS 850.4400*; USEPA: Washington, DC, USA, 2012. Available online: http://www.regulations.gov/document/EPA-HQ-OPPT-2009-0154-0027 (accessed on 15 January 2025).
27. Park, J.; Yoo, E.-J.; Shin, K.; Depuydt, S.; Li, W.; Appenroth, K.-J.; Lillicrap, A.D.; Xie, L.; Lee, H.; Kim, G.; et al. Interlaboratory validation of toxicity testing using the duckweed *Lemna minor* root-regrowth test. *Biology* **2022**, *11*, 37. [CrossRef] [PubMed]
28. *NY/T 3090-2017*; Chemical Pesticide-Guideline for *Lemma* sp. Growth Inhibition Test. Chinese Agriculture Industry Standard B17, Ministry of Agriculture Bulletin No. 2540. Field Test Asia: Singapore, 2017. Available online: https://www.chinesestandard.net/PDF/English.aspx/NYT3090-2017 (accessed on 15 January 2025).
29. Li, R.; Luo, C.; Qiu, J.; Li, Y.; Zhang, H.; Tan, H. Metabolomic and transcriptomic investigation of the mechanism involved in enantioselective toxicity of imazamox in *Lemna minor. J. Hazard. Mater.* **2022**, *425*, 127818. [CrossRef]

30. Naumann, B.; Eberius, M.; Appenroth, K.-J. Growth rate based dose-response relationships and EC-values of ten heavy metals using the duckweed growth inhibition test (ISO 20079) with *Lemna minor* L. *clone St. J. Plant Physiol.* **2007**, *164*, 1656–1664. [CrossRef] [PubMed]
31. EC Biological Test Method. *Test for Measuring the Inhibition of Growth Using the Freshwater Macrophyte, Lemna minor: Report EPS 1/RM/37*, 2nd ed.; Method Development and Applications Section, Environmental Technology Centre, Environment Canada: Ottawa, ON, Canada, 2007. Available online: https://publications.gc.ca/collections/collection_2013/ec/En49-7-1-37-eng.pdf (accessed on 15 January 2025).
32. Huebert, D.B.; Shay, J.M. Considerations in the assessment of toxicity using duckweeds. *Environ. Toxicol. Chem.* **1993**, *12*, 481–483. [CrossRef]
33. Olah, V.; Hepp, A.; Vaca, N.Y.G.; Tamas, M.; Meszaros, I. Retrospective analyses of archive phytotoxicity test data can help in assessing internal dynamics and stability of growth in laboratory duckweed cultures. *Aquat. Toxicol.* **2018**, *201*, 40–46. [CrossRef] [PubMed]
34. Taraldsen, J.E.; Norberg-King, T.J. New method for determining effluent toxicity using duckweed (*Lemna Minor*). *Environ. Toxicol. Chem.* **1990**, *9*, 761–767. [CrossRef]
35. Soukupova, I.; Beklova, M. Validation of duckweed microbiological test for assessing hazardous substances. *J. Biochem. Technol.* **2010**, *2*, S60–S61.
36. Kalcikova, G.; Marolt, G.; Kokali, A.J.; Gorvajn, A.Z. The use of multiwell culture plates in the duckweed toxicity test—A case study on Zn nanoparticles. *New Biotechnol.* **2018**, *47*, 67–72. [CrossRef]
37. Drobniewska, A.; Giebultowicz, J.; Wawryniuk, M.; Kierczak, P.; Nalecz-Jawecki, G. Toxicity and bioaccumulation of selected antidepressants in *Lemna minor* (L.). *Ecohydrol. Hydrobiol.* **2024**, *24*, 262–270. [CrossRef]
38. Michel, A.; Johnson, R.D.; Duke, S.O.; Scheffler, B.E. Dose-response relationships between herbicides with different modes of action and growth of *Lemna paucicostata*: An improved ecotoxicological method. *Environ. Toxicol. Chem.* **2004**, *23*, 1074–1079. [CrossRef]
39. Cayuela, M.L.; Millner, P.; Slovin, J.; Roig, A. Duckweed (*Lemna gibba*) growth inhibition bioassay for evaluating the toxicity of olive mill wastes before and during composting. *Chemosphere* **2007**, *68*, 1985–1991. [CrossRef] [PubMed]
40. Rivenbark, K.J.; Nikkhah, H.; Wang, M.; Beykal, B.; Phillips, T.D. Toxicity of representative organophosphate, organochlorine, phenylurea, dinitroaniline, carbamate, and viologen pesticides to the growth and survival of *H. vulgaris*, *L. minor*, and C. elegans. *Environ. Sci. Pollut. Res.* **2024**, *31*, 21781–21796. [CrossRef]
41. Shi, J.; Abid, A.D.; Kennedy, I.A.; Htristova, K.R.; Silk, W.K. To duckweeds (*Landoltia punctata*), nanoparticulate copper oxide is more inhibitory than the soluble copper in the bulk solution. *Environ. Pollut.* **2011**, *159*, 1277–1282. [CrossRef] [PubMed]
42. Alkimin, G.D.; Daniel, D.; Frankenbach, S.; Serodio, J.; Soares, A.M.V.M.; Barata, C.; Nunes, B. Evaluation of pharmaceutical toxic effects of non-standard endpoints on the macrophyte species *Lemna minor* and *Lemna gibba*. *Sci. Total Environ.* **2019**, *657*, 926–937. [CrossRef]
43. Irfan, M.; Meszaros, I.; Szabo, S.; Olah, V. Comparative phytotoxicity of metallic elements on duckweed *Lemna gibba* L. using growth- and chlorophyll fluorescence induction-based endpoints. *Plants* **2024**, *13*, 215. [CrossRef]
44. Brain, R.A.; Solomon, K.R. A protocol for conducting 7-day daily renewal tests with *Lemna gibba*. *Nat. Protoc.* **2007**, *2*, 979–987. [CrossRef] [PubMed]
45. Walbridge, C.T. *A Flow-Through Testing Procedure with Duckweed (Lemna minor L.)*; EPA-600/3-77-108; U.S. Environmental Protection Agency: Duluth, MN, USA, 1977. Available online: https://nepis.epa.gov/Exe/ZyPURL.cgi?Dockey=9101B3Q9.txt (accessed on 15 January 2025).
46. Davis, J.A. *Comparison of Static-Replacement and Flow-Through Bioassays Using Duckweed*; Lemna gibba G-2. EPA 560/6-81-003; U.S. Environmental Protection Agency, Office of Pesticides and Toxic Substances: Washington, DC, USA, 1981. Available online: https://nepis.epa.gov/Exe/ZyPURL.cgi?Dockey=9100AP3H.txt (accessed on 15 January 2025).
47. Coughlan, N.E.; Walsh, E.; Bolger, P.; Burnell, G.; O'Leary, N.; O'Mahoney, M.; Paolacci, S.; Wall, D.; Jansen, M.A.K. Duckweed bioreactors: Challenges and opportunities for large-scale indoor cultivation of Lemnaceae. *J. Clean. Prod.* **2022**, *336*, 130285. [CrossRef]
48. Wilson, P.C.; Hinz, F.O.; Farrell, I. Impacts of fulvic acid on the toxicity of the herbicide atrazine to *Lemna minor*. *Bull. Environ. Contam. Toxicol.* **2024**, *112*, 77. [CrossRef]
49. Kurnia, K.; Lin, Y.-T.; Farhan, A.; Malhotra, N.; Luong, C.T.; Hung, C.-H.; Roldan, M.J.M.; Tsao, C.-C.; Cheng, T.-S.; Hsiao, C.-D. Deep learning-based automatic duckweed counting using StarDist and its application on measuring growth inhibition potential of rare earth elements as contaminants of emerging concern. *Toxics* **2023**, *11*, 680. [CrossRef]
50. Okamura, H.; Togosmaa, L.; Sawamoto, T.; Fukushi, K.; Nishida, T.; Beppu, T. Effects of metal pyrithione antifoulants on freshwater macrophyte *Lemna gibba* G3 determined by image analysis. *Ecotoxicology* **2012**, *21*, 1102–1111. [CrossRef]
51. Cui, R.; Kwak, J.I.; An, Y.-J. Understanding boron toxicity in aquatic plants (*Salvinia natans* and *Lemna minor*) in the presence and absence of EDTA. *Aquat. Toxicol.* **2024**, *269*, 106886. [CrossRef] [PubMed]

52. Jansen, M.; Nagel, K. Useful methods (6): LemnaTec Scanalyzer, a useful tool for duckweed research and testing. *Duckweed Forum* **2016**, *4*, 282–286. Available online: http://www.ruduckweed.org/ (accessed on 15 January 2025).
53. Perera, W.H.; Meepagala, K.M.; Fronczek, F.R.; Cook, D.D.; Wedge, D.E.; Duke, S.O. Bioassay-guided isolation and structure elucidation of fungal and herbicidal compounds from *Ambrosia Salsola* (Asteraceae). *Molecules* **2019**, *24*, 835. [CrossRef]
54. Cruz, F.V.d.S.; Brant, H.S.C.; Ohlund, L.; Sleno, L.; Juneau, P. Tolerance and phytoremediation capacity of atrazine and S-metaloclor by two duckweeds. *Environ. Sci. Pollut. Res.* **2024**, *31*, 59382–59397. [CrossRef] [PubMed]
55. Adomas, B.; Sikorski, L.; Bes, A.; Warminski, K. Exposure of *Lemna minor* L. to gentian violet or Congo red is associated with changes in the biosynthesis pathway of biogenic amines. *Chemosphere* **2020**, *254*, 126752. [CrossRef]
56. Cox, K.L.; Manchego, J.; Meyers, B.C.; Czymmek, K.J.; Harkness, A. Automated imaging of duckweed growth and development. *Plant Direct* **2022**, *6*, 439. [CrossRef]
57. Olah, V.; Kosztanko, K.; Irfan, M.; Szabo, S.B.; Jansen, M.A.K.; Szabo, S.; Meszaros, I. Frond-level analyses reveal functional heterogeneity within heavy metal-treated duckweed colonies. *Plant Stress* **2024**, *11*, 100405. [CrossRef]
58. Lai, Y.-J.; Lu, P.-C.; Kung, Y. Duckweed-based optical biosensor for herbicide toxicity assessment. *Biosens. Bioelectron.* **2025**, *267*, 116739. [CrossRef] [PubMed]
59. Mazur, R.; Szoszkiewicz, K.; Lewicki, P.; Bedla, D. The use of computer image analysis in a *Lemna minor* L. bioassay. *Hydrobiologia* **2018**, *812*, 193–201. [CrossRef]
60. Haffner, O.; Kucera, E.; Drahos, P.; Ciganek, J.; Kozakova, A.; Urminska, B. *Lemna minor* bioassay evaluation using computer image analysis. *Water* **2020**, *12*, 2207. [CrossRef]
61. Subbaraman, B.; de Lange, O.; Ferguson, S.; Peek, N. The Duckbot: A system for automated imaging and manipulation of duckweed. *PLoS ONE* **2024**, *19*, e0296717. [CrossRef] [PubMed]
62. Bengtsson, B.-E.; Bongo, J.P.; Eklund, B. Assessment of duckweed *Lemna aequinoctialis* as a toxicological bioassay for tropical environments in developing countries. *Ambio* **1999**, *28*, 152–155. Available online: https://www.jstor.org/stable/4314867 (accessed on 19 December 2024).
63. Trenfield, M.A.; Harford, A.J.; Mooney, T.; Ellis, M.; Humphrey, C.; van Dam, R.A. 2019. Integrating laboratory and field studies to assess impacts of discharge from a uranium mine and validate a water quality guideline value for magnesium. *Integ. Environ. Assess. Manag.* **2019**, *15*, 64–76. [CrossRef]
64. Baudo, R.; Foudoulakis, M.; Arapi, G.; Perdaen, K.; Lanneau, W.; Paxinou, A.-C.M.; Kouvdou, S.; Persoone, G. History and sensitivity comparison of the *Spirodela polyrhiza* microbiotest and *Lemna* toxicity tests. *Knowl. Manag. Aquat. Ecosyst.* **2015**, *416*, 23. [CrossRef]
65. Zhao, X.-L.; Li, P.; Qu, C.; Lu, R.; Li, Z.-H. Phytotoxicity of environmental norfloxacin concentrations on the aquatic plant *Spirodela polyrrhiza*: Evaluation of growth parameters, photosynthetic toxicity and biochemical traits. *Compar. Biochem. Physiol. Part C* **2022**, *258*, 109365. [CrossRef]
66. Lalau, C.M.; Simioni, C.; Vicentini, D.; Ouriques, L.C.; Mohedano, R.A.; Puerari, R.C.; Matias, G. Toxicological effects of AgNPs on duckweed (*Landoltia punctata*). *Sci. Total Environ.* **2020**, *710*, 136318. [CrossRef]
67. Yang, G.-L.; Huang, M.J.; Tan, A.-J.; Lv, S.-M. 2021. Joint effects of naphthalene and microcystin-LR on physiological responses and toxin bioaccumulation of *Landoltia punctata*. *Aquat. Toxicol.* **2021**, *231*, 105710. [CrossRef] [PubMed]
68. Zou, X.; Li, P.; Huang, Q.; Zhang, H. The different response mechanisms of *Wolffia globosa*: Light-induce silver nanoparticle toxicity. *Aquat. Toxicol.* **2016**, *175*, 97–105. [CrossRef]
69. Antunes, P.M.C.; Kreager, N.J. Lead toxicity to *Lemna minor* predicted using a metal speciation chemistry approach. *Environ. Toxicol. Chem.* **2014**, *33*, 2225–2233. [CrossRef]
70. Gopalapillai, Y.; Vigneault, B.; Halo, B.A. Root length of aquatic plant *Lemna minor* L. as an optimal toxicity endpoint for biomonitoring of mining effluents. *IEAM (Integr. Environ. Assess. Manag.)* **2014**, *10*, 493–497. [CrossRef]
71. Xue, Y.; Wang, J.-q.; Huang, J.; Li, F.-y.; Wang, M. The response of duckweed (*Lemna minor* L.) roots to Cd and its chemical forms. *J. Chem.* **2018**, *2018*, 7274020. [CrossRef]
72. Park, A.; Kim, Y.-J.; Choi, E.-M.; Brown, M.T.; Han, T. A novel bioassay using root re-growth in *Lemna*. *Aquat. Toxicol.* **2013**, *140–141*, 415–424. [CrossRef]
73. Park, J.; Brown, M.T.; Depuydt, S.; Kim, J.K.; Won, D.-S.; Han, T. Comparing the acute sensitivity of growth and photosynthetic endpoints in three *Lemna* species exposed to four herbicides. *Environ. Pollut.* **2017**, *220*, 818–837. [CrossRef] [PubMed]
74. Liu, Y.; Zhao, X.; Ma, Y.; Dai, W.; Song, Z.; Wang, Y.; Shen, J.; He, X.; Yang, F.; Zhang, Z. Interaction of cerium oxide nanoparticles and ionic cerium with duckweed (*Lemna minor* L.): Uptake, distribution and phytotoxicity. *Nanomaterials* **2023**, *13*, 2523. [CrossRef] [PubMed]
75. Lee, H.; De Saeger, J.; Bae, S.; Kim, M.; Depuydt, S.; Heynderickx, P.M.; Wu, D.; Han, T.; Park, J. 2023. Giant duckweed (*Spirodela polyrhiza*) root growth as a simple and sensitive indicator of copper and chromium contamination. *Toxics* **2023**, *11*, 788. [CrossRef]

76. Rozman, U.; Turk, T.; Skalar, T.; Zupancic, M.; Korosi, N.C.; Marinsek, M.; Olivero-Verbel, J.; Kalcikova, G. An extensive characterization of various environmentally relevant microplastics—Material properties, leaching and ecotoxicity testing. *Sci. the Total Environ.* **2021**, *773*, 145576. [CrossRef]
77. Olah, V.; Hepp, A.; Mesaros, I. Comparative study on the sensitivity of turions and active fronds of giant duckweed (*Spirodela polyrhiza* (L.) Schleiden) to heavy metal treatments. *Chemosphere* **2015**, *132*, 40–46. [CrossRef]
78. Olah, V.; Hepp, A.; Meszaros, I. Assessment of Giant Duckweed (*Spirodela polyrhiza* L. Schleiden) turions as model objects in ecotoxicological applications. *Bull. Environ. Contam. Toxicol.* **2016**, *96*, 596–601. [CrossRef]
79. *ISO 4979:2023*; Water Quality—Aquatic Toxicity Test Based on Root Regrowth in Lemna Minor. International Organization for Standardization: Geneva, Switzerland, 2023. Available online: https://www.iso.org/standard/80751.html (accessed on 15 January 2025).
80. Ziegler, P. The developmental cycle of *Spirodela polyrhiza* turions: A model for turion-based duckweed overwintering? *Plants* **2024**, *13*, 2993. [CrossRef]
81. *ISO 20227:2017*; Water Quality—Determination of the Growth Inhibition Effects of Waste Waters, Natural Waters and Chemicals on the Duckweed *Spirodela polyrhiza*—Method Using a Stock Culture Independent Microbiotest. International Organization for Standardization: Geneva, Switzerland, 2017. Available online: https://www.iso.org/standard/67326.html (accessed on 15 January 2025).
82. Brain, R.A.; Cedergreen, N. Biomarkers in aquatic plants: Selection and utility. *Rev. Environ. Contam. Toxicol.* **2009**, *198*, 49–109. [CrossRef]
83. Pietrini, F.; Passatore, L.; Fischetti, E.; Carloni, S.; Ferrario, C.; Polesello, S.; Zacchini, M. Evaluation of morpho-physiological traits and contaminant accumulation ability in *Lemna minor* L. treated with increasing perfluorooctanoic acid (PFOA) concentrations under laboratory conditions. *Sci. Total Environ.* **2019**, *695*, 133828. [CrossRef]
84. Aliferis, K.A.; Materzok, S.; Paziotou, G.N.; Chrysayi-Tokousbalides, M. *Lemna minor* L. as a model organism for ecotoxicological studies performing ^{1}H NMR fingerprinting. *Chemosphere* **2009**, *76*, 967–975. [CrossRef] [PubMed]
85. Lomagin, A.G.; Ulyanova, L.V. A new bioassay on water pollution using duckweed *Lemna minor* L. *Soviet Plant Physiol.* **1993**, *40*, 137–138.
86. Yang, J.; Zhao, X.; Wang, X.; Xia, M.; Ba, S.; Lim, B.L.; Hou, H. Biomonitoring of heavy metals and their phytoremediation by duckweeds: Advances and prospects. *Environ. Res.* **2024**, *245*, 118015. [CrossRef]
87. Li, T.; Xiong, Z. A novel response of wild-type duckweed (*Lemna paucicostata* Hegelm.) to heavy metals. *Environ. Toxicol.* **2004**, *19*, 95–102. [CrossRef] [PubMed]
88. Henke, R.; Eberius, M.; Appenroth, K.-J. Induction of frond abscission by metals and other toxic compounds in *Lemna minor*. *Aquat. Toxicol.* **2011**, *101*, 261–265. [CrossRef]
89. Olah, V.; Hepp, A.; Lakatos, G.; Meszaros, I. Induced turion formation of *Spirodela polyrhiza* (L.) Schleiden. *Acta Biol. Szeged.* **2014**, *58*, 103–108. Available online: http://www2.sci.u-szeged.hu/ABS (accessed on 14 December 2024).
90. Appenroth, K.-J.; Krech, K.; Keresztes, A.; Fischer, W.; Koloczek, H. Effects of nickel on the chloroplasts of the duckweeds *Spirodela polyrhiza* and *Lemna minor* and their possible use in biomonitoring and phytoremediation. *Chemosphere* **2010**, *78*, 216–223. [CrossRef]
91. Sree, K.S.; Keresztes, A.; Mueller-Roeber, B.; Brandt, R.; Eberius, M.; Fischer, W.; Appenroth, K.-J. Phytotoxicity of cobalt ions on the duckweed *Lemna minor*—Morphology, ion uptake and starch accumulation. *Chemosphere* **2015**, *131*, 149–158. [CrossRef]
92. Bacıak, M.; Sikorski, L.; Piotrowicz-Cieslak, A.I.; Adomas, B. Content of biogenic amines in *Lemna minor* (common duckweed) growing in medium contaminated with tetracycline. *Aquat. Toxicol.* **2016**, *180*, 95–102. [CrossRef]
93. Elbasan, F.; Arikan-Abdulveli, B.; Ozfidan-Konakci, C.; Yildiztugay, E.; Tarhan, I.; Celik, B. Exploring the defense strategies of benzalkonium chloride exposure on the antioxidant system, photosynthesis and ROS accumulation in *Lemna minor*. *Chemosphere* **2024**, *363*, 142924. [CrossRef] [PubMed]
94. Olah, V.; Hepp, A.; Irfan, M.; Meszaros, I. Chlorophyll fluorescence imaging-based duckweed phenotyping to assess acute phytotoxic effects. *Plants* **2021**, *10*, 2763. [CrossRef]
95. Küster, A.; Altenburger, R. Development and validation of a new fluorescence-based bioassay for aquatic macrophyte species. *Chemosphere* **2007**, *67*, 19–201. [CrossRef]
96. Küster, A.; Pohl, K.; Altenburger, R. A fluorescence-based bioassay for aquatic macrophytes and its suitability for effect analysis of non-photosystem II inhibitors. *Environ. Sci. Pollut. Res.* **2007**, *14*, 377–383. [CrossRef]
97. Pietrini, F.; Passatore, L.; Carloni, S.; Zacchini, M. Non-standard physiological endpoints to evaluate the toxicity of emerging contaminants in aquatic plants: A case study on the exposure of *Lemna minor* and *Spirodela polyrhiza* (L.) Schleid. to dimethyl phthalate (DMP). In *Emerging Contaminants and Plants. Emerging Contaminants and Associated Treatment Technologies*; Aftab, T., Ed.; Springer: Cham, Switzerland, 2023. [CrossRef]
98. Razinger, J.; Drinovec, L.; Zrimec, A. Real-time in vivo visualization of oxidative stress in duckweed (*Lemna minor* L.). *Cent. Eur. J. Biol.* **2007**, *3*, 351–363. [CrossRef]

99. Persic, V.; Derd, T.; Varga, M.; Hackenberger, B.K. Real-time CO_2 uptake/emission measurements as a tool for early indication of toxicity in Lemna-tests. *Aquat. Toxicol.* **2019**, *206*, 154–163. [CrossRef] [PubMed]
100. Kumar, S.; Barth, A. Following enzyme activity with infrared spectroscopy. *Sensors* **2010**, *10*, 2626–2637. [CrossRef]
101. Schwaighofer, A.; Akhgar, C.K.; Lendl, B. Broadband laser-based mid-IR spectroscopy for analysis of proteins and monitoring of enzyme activity. *SAA* **2021**, *253*, 119563. [CrossRef]
102. Pietrini, F.; Zacchini, M. A new ecotoxicity assay for aquatic plants; Eco-Tox Photosystem Tool (ETPT). *Trends Plant Sci.* **2020**, *25*, 1266–1267. [CrossRef]
103. Wang, W.; Li, R.; Zhu, Q.; Tang, X.; Zhao, Q. Transcriptomic and physiological analysis of common duckweed *Lemna minor* responses to NH_4^+ toxicity. *BMC Plant Biol.* **2016**, *16*, 92. [CrossRef]
104. Xu, H.; Yu, C.; Xia, X.; Li, M.; Li, H.; Wang, Y.; Wang, S.; Wang, C.; Ma, Y.; Zhou, G. Comparative transcriptome analysis of duckweed (*Landoltia punctata*) in response to cadmium provides insights into molecular mechanisms underlying hyperaccumulation. *Chemosphere* **2018**, *190*, 154–165. [CrossRef] [PubMed]
105. Muthan, B.; Wang, J.; Welti, R.; Kosma, D.K.; Yu, L.; Deo, B.; Khatiwada, S.; Vulavala, V.K.R.; Childs, K.L.; Xu, C.; et al. Mechanisms of *Spirodela polyrhiza* tolerance to FGD wastewater-induced heavy-metal stress: Lipidomics, transcriptomics, and functional validation. *J. Hazard. Mater.* **2024**, *469*, 133951. [CrossRef]
106. Loll, A.; Reinwald, H.; Ayobahan, S.U.; Göckener, B.; Salinas, G.; Schäfers, C.; Schlich, K.; Hamscher, G.; Eilebrecht, S. Short-term test for toxicogenomic analysis of ecotoxic modes of action in *Lemna minor*. *Environ. Sci. Technol.* **2022**, *56*, 11504–11515. [CrossRef]
107. Su, C.; Jiang, Y.; Yang, Y.; Zhang, W.; Xu, Q. Responses of duckweed (*Lemna minor* L.) to aluminum stress: Physiological and proteomics analyses. *Ecotoxicol. Environ. Saf.* **2019**, *170*, 127–140. [CrossRef]
108. Kostopoulou, S.; Ntatsi, G.; Arapis, G.; Aliferis, K.A. Assessment of the effects of metribuzin, glyphosate, and their mixtures on the metabolism of the model plant *Lemna minor* L. applying metabolomics. *Chemosphere* **2020**, *239*, 124582. [CrossRef]
109. Wahman, R.; Cruzeiro, C.; Graßmann, J.; Schröder, P.; Letzel, T. The changes in *Lemna minor* metabolomic profile: A response to diclofenac incubation. *Chemosphere* **2022**, *287*, 132078. [CrossRef] [PubMed]
110. Hu, L.-X.; Ying, G.-G.; Chen, X.-W.; Huang, G.-Y.; Liu, Y.-S.; Jiang, Y.-X.; Pan, C.-G.; Tian, F.; Martin, F.L. Fourier transform infrared spectroscopy as a novel approach to providing effect-based endpoints in duckweed toxicity testing. *Environ. Toxicol. Chem.* **2017**, *36*, 346–353. [CrossRef] [PubMed]
111. Wu, Y.-L.; Xiong, Q.; Wang, B.; Liu, Y.-S.; Zou, P.-L.; Hu, L.-X.; Liu, F.; Ying, G.-G. Screening of structural and functional alterations in duckweeds (*Lemna minor*) induced by per- and polyfluoroalkyl substances (PFASs) with FTIR spectroscopy. *Environ. Pollut.* **2023**, *317*, 120671. [CrossRef]
112. Zhang, T.; Lu, Q.; Su, C.; Yang, Y.; Hu, D.; Xu, Q. Mercury induced oxidative stress, DNA damage, and activation of antioxidative system and Hsp70 induction in duckweed (*Lemna minor*). *Ecotoxicol Environ. Saf.* **2017**, *143*, 46–56. [CrossRef]
113. Brkanac, S.R.; Domijan, A.-M.; Stefanic, P.P.; Maldini, K.; Sikiric, M.D.; Bol, V.V.; Cvjetko, P. Difference in the toxic effects of micro and nano ZnO particles on *L. minor*—An integrative approach. *Environ. Sci. Pollut. Res.* **2024**, *31*, 58830–58843. [CrossRef]
114. Pietrini, F.; Iannilli, V.; Passatore, L.; Carloni, S.; Sciacca, G.; Cerasa, M.M.; Zacchini, M. Ecotoxicological and genotoxic effects of dimethyl phthalate (DMP) on *Lemna minor* L. and *Spirodela polyrhiza* (L.) Schleid. plants under a short-term laboratory assay. *Sci. Total Environ.* **2022**, *806*, 150972. [CrossRef] [PubMed]
115. Ma, X.; Jiang, Y.; Qu, Z.; Yang, Y.; Wang, W.; He, Y.; Yu, Y.; Luo, X.; Liu, W.; Han, W.; et al. Overexpression of phosphoserine aminotransferase (*PSAT*)-enhanced cadmium resistance and accumulation in duckweed (*Lemna turionifera* 5511). *Plants* **2024**, *13*, 627. [CrossRef] [PubMed]
116. Van Hoeck, A.; Horemans, N.; Nauts, R.; VanHees, M.; Vandenhove, H.; Blust, R. *Lemna minor* plants chronically exposed to ionizing radiation: RNA-seq analysis indicates a close rate dependent shift from acclimation to survival strategies. *Plant Sci.* **2017**, *257*, 84–95. [CrossRef]
117. Fu, L.; Ding, Z.; Sun, X.; Zhang, J. Physiological and transcriptomic analysis reveals distorted ion homeostasis and response in the freshwater plant *Spirodela polyrhiza* L. under salt stress. *Genes* **2019**, *10*, 743. [CrossRef] [PubMed]
118. Shang, S.; Zhang, Z.; Li, L.; Chen, J.; Zang, Y.; Liu, X.; Wang, J.; Tang, X. Transcriptome analysis reveals gene expression patterns of *Spirodela polyrhiza* response to heat stress. *Int. J. Biol. Macromol.* **2023**, *225*, 767–775. [CrossRef] [PubMed]
119. Yu, C.; Zhao, X.; Qi, G.; Bai, Z.; Wang, Y.; Wang, S.; Ma, Y.; Liu, Q.; Hu, R.; Zhou, G. Integrated analysis of transcriptome and metabolites reveals an essential role of metabolic flux in starch accumulation under nitrogen starvation in duckweed. *Biotechnol. Biofuels* **2017**, *10*, 167. [CrossRef]
120. Fu, L.; Tan, D.; Sun, X.; Ding, Z.; Zhang, J. Transcriptional analysis reveals potential genes and regulatory networks involved in salicylic acid-induced flowering in duckweed (*Lemna gibba*). *Plant. Physiol. Biochem.* **2020**, *155*, 512–522. [CrossRef]
121. Wang, W.; Wu, Y.; Messing, J. RNA-Seq transcriptome analysis of Spirodela dormancy without reproduction. *BMC Genom.* **2014**, *15*, 60. [CrossRef]

122. Pasaribu, B.; Acosta, K.; Aylward, A.; Liang, Y.; Abramson, B.W.; Colt, K.; Hartwick, N.T.; Shanklin, J.; Michael, T.P.; Lam, E. Genomics of turions from the Greater Duckweed reveal its pathways for dormancy and re-emergence strategy. *New Phytol.* **2023**, *239*, 116–131. [CrossRef]
123. Wang, W.; Yang, C.; Tang, X.; Gu, X.; Zhu, Q.; Pan, K.; Hu, Q.; Ma, D. Effects of high ammonium level on biomass accumulation of common duckweed *Lemna minor* L. *Environ. Sci. Pollut. Res.* **2014**, *21*, 14202–14210. [CrossRef]
124. Afshana; Dyar, M.A.; Reshi, Z.A. Induced genotoxicity and oxidative stress in plants: An overview. In *Induced Genotoxicity and Oxidative Stress in Plants*; Khan, Z., Ansari, M.Y.K., Shahwar, D., Eds.; Springer Nature: Singapore, 2021; pp. 1–27. [CrossRef]
125. Picinini-Zambelli, J.; Garcia, A.L.H.; Da Silva, J. Emerging pollutants in the aquatic environment: A review of genotoxic impacts. *MRR* **2025**, *795*, 108519. [CrossRef] [PubMed]
126. Kaur, L.; Kanwar, N. Duckweed: A model for phytoremediation technology. *Holist. Approach Environ.* **2022**, *12*, 39–58. Available online: https://casopis.hrcpo.com/volume-12-issue-1-kaur-et-al/ (accessed on 18 January 2025). [CrossRef]
127. Ekperusi, A.O.; Sikoki, F.D.; Nwachukwu, E.O. Application of common duckweed (*Lemna minor*) in phytoremediation of chemicals in the environment; State and future perspective. *Chemosphere* **2019**, *223*, 285–309. [CrossRef] [PubMed]
128. Ekperusi, A.O.; Sikoki, F.D.; Nwachukwu, E.O. Sorption of cadmium, chromium, lead, and vanadium from artificial wetlands using *Lemna aequinoctialis*. *Int. J. Phytoremed.* **2023**, *26*, 873–881. [CrossRef] [PubMed]
129. Yang, L.; Yao, J.; Sun, J.; Shi, L.; Cen, Y.; Su, J. The Ca^{2+} signaling, Glu, and GABA responds to Cd stress in duckweed. *Aquat. Toxicol.* **2020**, *218*, 105352. [CrossRef] [PubMed]
130. Yang, L.; Han, H.; Liu, M.; Zuo, Z.; Zhou, K.; Lü, J.; Zhu, Y.; Bai, Y.; Wang, Y. Overexpression of the *Arabidopsis* photorespiratory pathway gene, serine: Glyoxylate aminotransferase (*AtAGT1*), leads to salt stress tolerance in transgenic duckweed (*Lemna minor*). *Plant Cell. Tiss. Organ Cult.* **2013**, *113*, 407–416. [CrossRef]
131. Canto-Pastor, A.; Molla-Morales, A.; Ernst, E.; Dahl, W.; Zhai, J.; Yan, Y.; Meyers, B.C.; Shanklin, J.; Martienssen, R. Efficient transformation and artificial miRNA gene silencing in *Lemna minor*. *Plant Biol.* **2015**, *17* (Suppl. S1), 59–65. [CrossRef]
132. Gan, W.C.; Ling, A.P.K. CRISPR/Cas9 in plant biotechnology: Applications and challenges. *BioTechnologia* **2022**, *103*, 81–93. [CrossRef]
133. Liu, Y.; Wang, Y.; Xu, S.; Tang, X.; Zhao, J.; Yu, C.; He, G.; Xu, H.; Wang, S.; Tang, Y.; et al. Efficient genetic transformation and CRISPR/Cas9-mediated genomic editing in *Lemna aequinoctialis*. *Plant Biotechnol. J.* **2019**, *17*, 2143–2152. [CrossRef]

MDPI AG
Grosspeteranlage 5
4052 Basel
Switzerland
Tel.: +41 61 683 77 34

Journal of Xenobiotics Editorial Office
E-mail: jox@mdpi.com
www.mdpi.com/journal/jox

www.ingramcontent.com/pod-product-compliance
Lightning Source LLC
Chambersburg PA
CBHW061629210926
51850CB00009B/2490
* 9 7 8 3 7 2 5 8 4 8 1 3 3 *